L4643

LED

AF795128

505006001

Machine-tool dynamics
An introduction

Machine-tool dynamics
An introduction

D. B. WELBOURN M.A., C.Eng., F.I.Mech.E., F.I.E.E., Member Verein Deutscher Ingenieure

and

J. D. SMITH M.A., Ph.D., C.Eng., M.I.Mech.E.

University Lecturers in Engineering and
Fellows of Selwyn College, Cambridge

CAMBRIDGE

at the University Press 1970

Published by the Syndics of the Cambridge University Press
Bentley House, 200 Euston Road, London, N.W.1
American Branch: 32 East 57th Street, New York, N.Y. 10022

© Cambridge University Press 1970

Library of Congress Catalogue Card Number: 71-101447
Standard Book Number: 521 07765 6

50060

2 · IX · 70 .

SHEFFIELD UNIV.
APPLIED SCIENCE
LIBRARY.

Printed in Great Britain
at the University Press, Aberdeen

Contents

Preface

The object of this book is to provide the reader with the sort of introduction to the dynamics of machine tools which I wish that I could have found in 1965 when I was first asked to look into some of the problems of gear cutting machinery by W. E. Sykes Ltd; to that firm, and to its Engineering Director, Mr J. C. Crockett, I am much indebted not merely for many interesting discussions but for the loan of machines and equipment to the Cambridge University Engineering Laboratory. Since then work on gear cutting and shaving has been pursued with help from Churchill Gear Machines Ltd, and the Drummond Division of Staveley Machine Tools Ltd, both of whom have loaned machines and given generous help through members of their staffs. For the past three years the work has been financed by a grant from the Science Research Council. To all of these organizations, and to many others who have collaborated with us more recently, Dr Smith and I are much indebted.

Although certain sections of this book, and in particular the whole of Chap. 10, were written by Dr Smith, the shape of it is mine. I came to machine tools with a background in control theory and very soon decided that they could most profitably be looked at from this standpoint. The analytical approach in this book is therefore entirely uniform with that which I used in a previous book *The Essentials of Control Theory for Mechanical Engineers* (Arnold 1963). Perhaps the most useful feature of this present work is not that it contains much that is new, but that it handles the various sections in a uniform manner and that it is written at a point in time when I am still very clear about the nature of the problems which most puzzled me when I entered this field; it is intended for the beginner rather than for the expert. I hope however that even the expert may find some helpful ideas. No problem is discussed which I have not faced, nor theoretical treatment given which I have not used to solve urgent industrial problems.

We have attempted to simplify the writing of equations by writing D for $\frac{d}{dt}$; and where some variable has a mean value, e.g. F, we have written it as a capital letter and then written small instantaneous changes from the mean in lower case, e.g. f. The total value at any instant is therefore $F+f$.

This book is an expansion by Dr J. D. Smith and myself of a set of lectures written by me for the Escuela Tecnica Superior de Ingenieros Industriales, Barcelona. I am deeply indebted to Dr Smith, who has brought his great experience and insight into machine tools to bear in helping to write this book, and to widening my knowledge and understanding of the subject. We received many helpful suggestions from our colleague Cdr A. W. Nutbourne R.N. (rt.) who read the whole book in draft.

We also owe much to various research students who have worked on different aspects of the problems of shaping, hobbing and shaving machines; B. L. Tramontini, D. A. D. Cooke, J. A. Ritchie and J. E. Sharpe.

We are grateful to Mrs I. U. Maclaren and Mrs Dorothy Ward for their skill and patience in the preparation of the typescript.

D. B. WELBOURN

February 1969

1. The machine tool as a closed loop

1. The main theme of this book is that a machine tool will deflect if a force is applied between the workpiece and the cutting tool. Typically the static stiffness will be found to be of the order of 500 000 lbf/in. Since the machine has distributed mass, it will have a number of resonant frequencies. These may be excited in obvious manners by milling cutters or hobs which cut intermittently at high frequencies; but they may also be excited by apparently continuous cutting processes such as turning, boring and planing. The reason for this lies in the fact that any change in force due to a hard spot in the material, or material clearing itself from a built-up edge on the cutting tool, will cause a change in the amount that the machine deflects, and hence in the depth of cut. This will lead to a further variation in the cutting force and so on. Sometimes this process builds up into the vibratory condition known as chatter; sometimes it dies away again. The physical principles involved are now well understood, and these lead to an understanding of how to avoid trouble. For many reasons the exact conditions under which chatter will occur still cannot be predicted with certainty at the design stage of a new machine; but given a proper understanding of the problems involved, the designer can do much to reduce the areas of uncertainty in the work which he has to do. In particular, he can learn how to increase both the stiffness of machines, and also the amount of damping in them; much of this book is concerned with these two problems.

2. A machine tool must be statically stiff if it is to produce accurate work: but static stiffness is only a particular case of dynamic stiffness: in this book machine tools, and the cutting processes which take place in them, will mainly be considered dynamically. In particular, attention will be paid to the causes of chatter, since its investigation throws light on the desired state of affairs when chatter is not obviously present.

3. For purposes both of analytical and also of experimental investigation the machine tool and the cutting process may be regarded as two distinct elements which form a closed loop typical of those found in control systems: the output from one is the input to the other. If each part of this loop, shown in Fig. 1.1, can be described, then the behaviour of the machine as a production unit can be described, and its liability to chatter assessed. This loop is concerned with variations from some desired mean steady state: the fundamental error in much published work in this field is the assumption that steady state cutting occurs. No material is perfectly homogeneous, and inevitably small variations occur randomly and frequently in the cutting process.

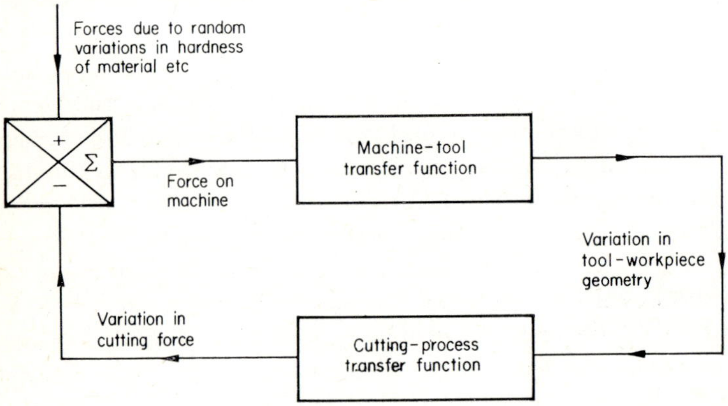

Fig. 1.1. Closed loop for machine tool.

4. The analysis in this book is largely linear, although most machine tools are very non-linear in their responses; and so are cutting processes. Despite this fact, many useful inferences can be drawn from linear theory so long as its limitations are always kept in mind. The basic assumption is that a curve may be approximated at any point by its tangent; the methods of analysis are those of classical control theory.

5. Recently a new theoretical approach, taken also from control theory, has started to be applied. This concerns the use of statistical control theory, and correlation techniques; given adequate computer facilities, it is valuable in enabling an analysis to be performed of a machine under normal operating conditions. The statistical approach will not be used in this book, since the minimum mathematics required lies outside most undergraduate courses.

6. The central chapter in this book is Chap. 5 in which some of the mechanisms of chatter are discussed. In order to appreciate fully the assumptions used in it, the previous chapters on the machine-tool frame and the cutting process need to be understood; but these chapters may seem dry and pedantic until their use is seen. This is an organizational problem of any textbook and the reader is therefore advised to skim through this book as a whole before he (or she) settles down to study the earlier chapters in detail. Their purpose should then be clearer.

7. Problem

1. Redraw Fig. 1.1 for a copy lathe with an input of desired shape fitted with in-process gauging.

2. Machine-tool frames

1. Static deflections. Machine-tool frames are usually made of cast iron or of fabricated mild steel, and models of them in Perspex. The most important data for these materials is given below:

	Mild steel	Cast iron	Perspex	
E	30×10^{6}psi	$16{\cdot}5\times10^{6}$psi	$4{\cdot}5\times10^{5}$psi	Young's Modulus
G	$11{\cdot}5\times10^{6}$psi	$6{\cdot}5\times10^{6}$psi	$1{\cdot}67\times10^{5}$psi	Modulus of Rigidity
ρ	0·281 lb/in^3	0·281 lb/in^3	0·043 lb/in^3	Density
ν	0·25	0·29	0·35	Poisson's Ratio
α	11×10^{-6}/°C	9×10^{-6}/°C	9×10^{-5}/°C	Coefficient of Expansion

Weight for weight a mild steel fabrication may be expected to be about twice as stiff as a similar design in ordinary cast iron.

The figures for cast iron need to be taken with considerable caution since the value of Young's Modulus is dependant on the load applied, and also on the grade of iron. At zero stress for an iron with Ultimate Tensile Strength (UTS) of 10 ton/in^2 the modulus will be about 15×10^6 psi, whereas for an iron with an UTS of 26 ton/in^2 the modulus will be about 21×10^6 psi. For the lower grade iron the modulus will decrease by about $0{\cdot}8\times10^6$ psi for each 1 ton/in^2 applied, whereas for the high grade iron the decrease will only be about $0{\cdot}3\times10^6$ psi. The stress/strain curve is non-linear and is also hysteretic.

The figures for Perspex also need interpreting with caution, since at 100 Hz (cycles per second) the Young's Modulus rises to about 7×10^5 psi.

2. Under cutting forces, deflections which may be considerable will occur in the frame. A well designed machine tool may have a stiffness of about 500 000 lbf/in between the workpiece and the tool, but very much lower values are common. Since cutting forces of a ton are not uncommon, deflections of 0·004 in may be expected even in favourable circumstances; and much bigger ones are frequent. This will affect the accuracy of the work produced even if no dynamic effects occur.

3. A quick estimate of the stiffness of a machine tool may often be obtained with the aid of a proving ring and a clockgauge, the load being applied either by means of a small jack, or the feed mechanism of the machine itself. A typical set of curves obtained by this method are shown in Fig. 2.1. These show typical hysteresis loops due to slip in nominally rigid joints. That this occurs may be demonstrated not merely by measurement but also by the fact that on dismantling a machine, fretting corrosion will frequently be found to be present on joint faces.

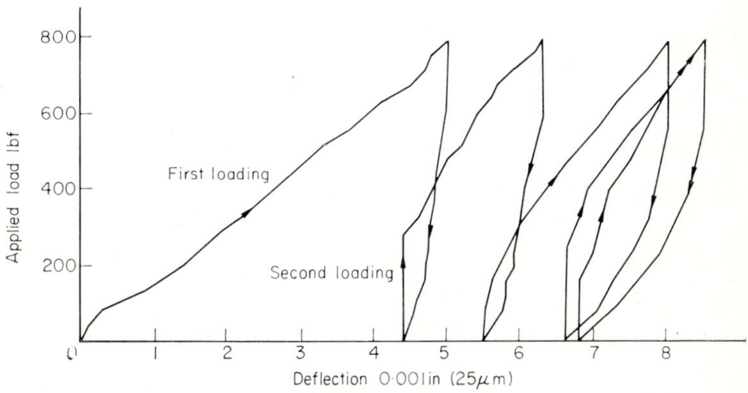

Fig. 2.1. Static loading hysteresis curves.

The surprising shape of this repeated loading curve cannot easily be explained during the first few cycles. When the curve has settled down to a repeatable hysteresis loop the explanation of its shape is straightforward. In this case the bolts give rise to a frictional force between the faces corresponding to about 100 lbf on the vertical scale. When the load is applied externally to the system, no deflection occurs until the frictional force is overcome. The load is then carried by the bolts assisted by friction until reversal occurs. Due to the fact that the direction of the frictional force has now reversed, no movement occurs until the external applied force has dropped by twice the amount of the frictional force; only then can the bolts start to straighten once more.

4. Cross-receptances. Due to the fact that most frames are asymmetrical, pushing them in one direction will in general cause them to move in some direction not along the line of action of the force. The simplest example of this is to be found when an **I**-beam is loaded about an axis which is not one of its principal axes. To obtain the deflection the force must be resolved perpendicular to the principal axes, and the vector sum of the deflections added: see Fig. 2.2.

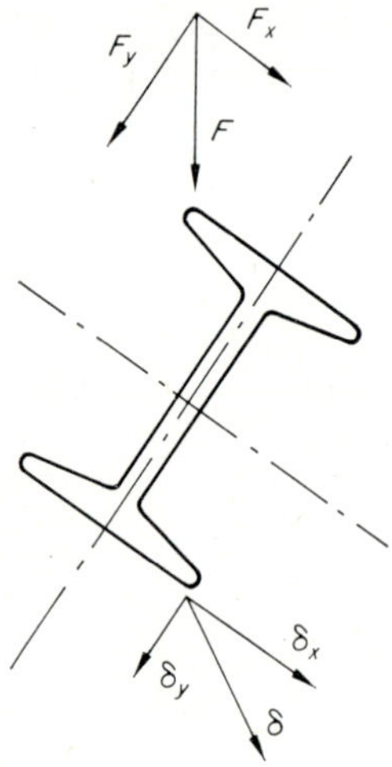

Fig. 2. 2. Cross receptance effect.

The movement in a given direction per unit force is known as the *receptance or flexibility* of a system (it is of course the inverse of the stiffness) and the movements in directions at right angles to the line of action of the force are known as cross-receptances. Their directions must be carefully specified for some consistent set of axes.

The concept of cross-receptance is a useful one in linear systems, since if the cross-receptances are known, and the forces also, the total deflections of a system may be calculated. This is however not true in non-linear systems, and most machine tools are non-linear. To obtain results useful from a design point of view, the line of action of the cutting force must be known, the machine loaded in this direction, and the loading must be at the actual force level which is normally met. The practice of deducing the effects of large loads from those of small ones can be most misleading.

5. Weak places in frames. Measurements frequently show that a large proportion of the flexibility of a frame is concentrated at a few places in the design, and that large changes in the stiffness can be effected by relatively small changes in design. Holes and small cut-outs can have surprisingly large effects on overall stiffness.

6. Model making. The best way at the design stage in which to find weak places is to test a simple model of the machine frame. The simplest materials are cartridge paper and glue; cardboard and glue yield more information, and can also be used for dynamic testing, but once they have failed due to buckling they are useless.

Fig. 2.3. Foam model.

The best material, however, is flexible plastic foam which may be bought cheaply in sheets of varying thickness, and may be cut up quickly and cheaply by means of a red-hot wire. The material is easy to stick together, and the models, Fig. 2.3, can yield a large amount of information very quickly.

7. Thermal distortion. Large distortions often occur in machine tools due to thermal distortion caused either by the change in temperature of the workshop, or by the change in temperature of the lubricant, coolant or cutting oil in the machine.

Change in temperature of the workshop, unless it is temperature controlled, can lead to large cyclical errors when cutting big gears, since the more massive parts of the machine will warm up and cool down much more slowly than the smaller components, thus giving rise to differential expansion. The effect on cumulative pitch accuracy may readily be measured on a large gear, which may spend a week or more on the table, particularly in late spring, when the nights may be very cold and the days very hot. The assembly and fitting of large machine tools may be greatly speeded up by carrying it out in a temperature controlled shop.

Changes of temperature in the lubricant and cutting oils may cause considerable distortions in small machine tools, particularly if they are operating on an intermittent cycle. Since for both cast iron and mild steel the coefficient of thermal expansion is of the order of $10^{-5}/°C$, a change of temperature of 10°C will cause a change in length of 0·001 in in 10 in. Particular care must be taken to make certain that coolant tanks are symmetrically disposed within the frame of a machine, or that they are separated from it entirely, in such a manner that they cannot transmit appreciable amounts of heat to one portion of the frame.

8. Dynamic behaviour. The densities ρ of both cast iron and mild steel are about the same (0·28 lb/in^3), but their elastic moduli in bending and shear are in the ratio 1:2. Since the resonant frequency of a beam in bending depends on $\sqrt{\frac{E}{\rho}}$ and in torsion on $\sqrt{\frac{G}{\rho}}$, changing a design from cast iron to mild steel for similar volumes of metal may be expected to raise all resonant frequencies by a factor of about $\sqrt{2}$.

9. Damping capacity. While the damping capacity of cast iron is very much greater than that of mild steel, the fact is irrelevant in most machine design, since most of the damping occurs in joints, bearings, and rubbing surfaces. Changing from a cast iron design to a welded mild steel design often results in the amount of inherent damping in the system being reduced due to elimination of bolted joints. If this results in too little damping, then vibration dampers must be added.

10. Dynamic testing of machines. When a machine tool is chattering, the force pattern repeats itself at frequencies between about 20 and 200 Hz. Although these patterns are not sinusoidal, they may be broken down into a series of sine waves using Fourier analysis. The most convenient way of testing a machine dynamically is therefore to excite it sinusoidally over a range of frequencies. While a lot of useful work can be done with portable electromagnetic exciters at low force levels of the order of ± 45 lbf, the only safe way of testing a machine is by exciting it at the force levels which occur in practice and directing these along the line of the known resultant force which occurs. For this purpose hydraulic vibrators, which give a sinusoidal force superimposed on a static

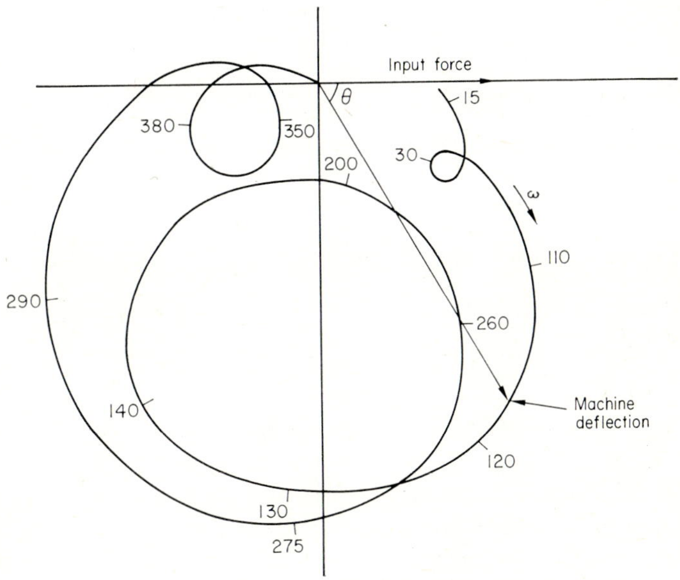

Fig. 2.4. Harmonic response locus. θ is angle of lag.

mean force, are normally necessary. In many machines great difficulties exist in obtaining fully representative results due to the difficulties of applying oscillatory forces when the feed mechanisms are working, but the errors due to testing a machine at rest generally appear to be small if high force levels are used.

A typical response curve for a machine tool is shown in Fig. 2.4 which shows the harmonic response locus. The reasons for its shape are:

a. Distributed spring-mass systems tend to give rise to circles on this diagram.

b. Spring-mass systems with hysteretic damping, which is a good rough representation of many machine-tool systems, also tend to give circles with a segment missing at low frequencies.

c. Coupled vibrating systems tend to give rise to curves with loops in them.

While harmonic response diagrams are useful in that they show the phase as well as the amplitude of the response, they do not make the amplitude relationships so clear as the more conventional amplitude-frequency form of plot shown in Fig. 2.5.

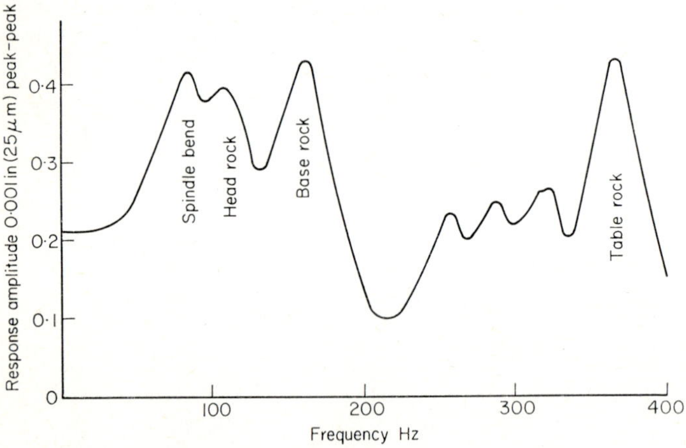

Fig. 2.5. Amplitude response curve.

11. Modal shapes. The amplitudes and phases of all parts of a machine vibrating at a resonant frequency may be measured by use of pick-ups, and plotted to show, on an exaggerated scale, the way in which the machine deforms. The resultant shape is termed the mode shape. In general, different mode shapes will be found for each resonant frequency, and are often strikingly different from one another for differing frequencies. These mode shapes, such as that shown in Fig. 2.6, are of great use in helping to determine the weak places in a structure, since discontinuities in the slopes of the lines indicate potential trouble spots.

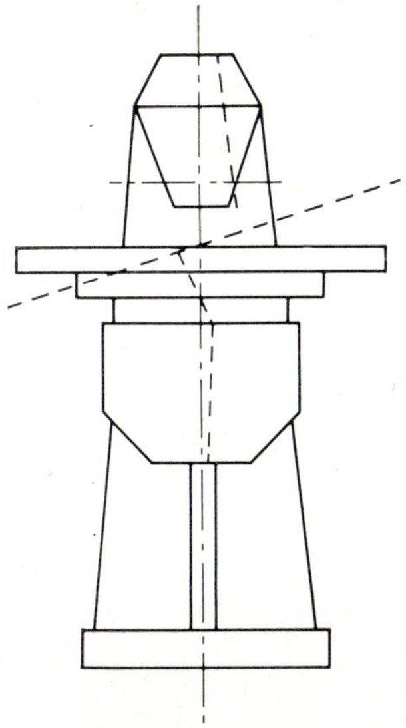

Fig. 2.6. Mode shape. Deflected position at resonance shown – – – –.

12. Equivalent dynamic system for machine. No difficulties are likely to occur with a machine from the vibration point of view unless the machine is being excited very close to one of its resonant frequencies. For most purposes the machine may then be represented adequately by a simple damped spring-mass system whose undamped natural frequency may be taken as the resonant frequency. The damping may be assumed to be either viscous or hysteretic* as may be found to fit the measurements better.

13. Foundations. Foundations also represent a damped spring-mass system, sometimes of a complicated nature, and the apparent resonant behaviour of a machine may vary greatly from one set

* Hysteretic damping approximates to that provided by Coulomb friction: but see Bibliography.

of foundations to another, or with different forms of fixing to the same foundations. In addition, foundations move due to the fact that concrete takes many months to cure completely, due to settling, and due to the fact that the subsoil normally moves with the amount of moisture in it. This can be particularly serious near to rivers and harbours, and can result in varying amounts of distortion in the machine and variations in the accuracy of the work produced. The state of the tide has been measurable on some marine gear-hobbing machines!

14. Vibration testing of models of machine-tool frames. Estimating the behaviour of a frame from the drawing of the machine is very difficult when designing a new machine, and models should normally be made and tested. Cardboard models are sometimes

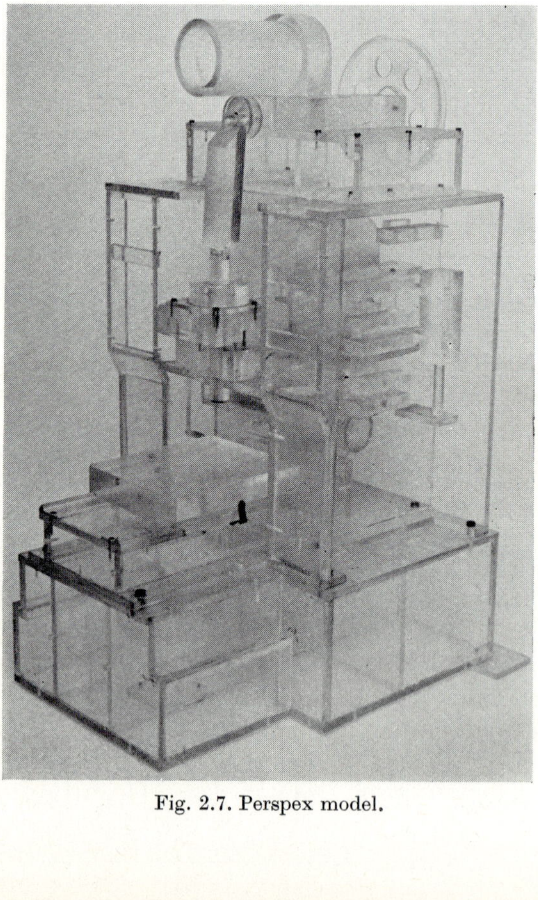

Fig. 2.7. Perspex model.

used, but Perspex models are better, and have the advantage that they are transparent. Even a relatively simple one such as that shown in Fig. 2.7 may cost (at 1968 prices) £40 in material and £75 in direct labour without overheads; but this may be money well spent. Perspex models are less satisfactory for static testing, because the material creeps under load. This may be overcome by using a Perspex proving ring to apply the load.

When testing dynamically to investigate possible resonant frequencies, if

ω is the angular frequency

L a typical dimension

and the subscripts m and f are used for the *m*odel and *f*ull scale machine respectively,

to satisfy dynamic similarity,

$$\frac{\omega_f^2 L_f^2 \rho_f}{E_f} = \frac{\omega_m^2 L_m^2 \rho_m}{E_m}$$

the expression being non-dimensional. The same expressions hold when G is substituted for E. Rearranging,

$$\frac{\omega_f}{\omega_m} = \frac{L_m}{L_f}\sqrt{\frac{\rho_m}{\rho_f}\frac{E_f}{E_m}}$$

For Steel Machine and Perspex Model

$$\frac{\omega_f}{\omega_m} = 3{\cdot}2\,\frac{L_m}{L_f} \qquad \text{when } E_m \text{ for Perspex is } 0{\cdot}45\times10^6 \text{ psi}$$

and

$$2{\cdot}56\,\frac{L_m}{L_f} \qquad \text{when } E_m \text{ for Perspex is } 0{\cdot}70\times10^6 \text{ psi}$$

For Cast Iron ($E = 16{\cdot}5\times10^6$ psi) and Perspex Model

$$\frac{\omega_f}{\omega_m} = 2{\cdot}37\,\frac{L_m}{L_f} \qquad \text{when } E_m \text{ for Perspex is } 0{\cdot}45\times10^6 \text{ psi}$$

and

$$1{\cdot}9\,\frac{L_m}{L_f} \qquad \text{when } E_m \text{ for Perspex is } 0{\cdot}70\times10^6 \text{ psi.}$$

As examples, a resonance of 100 Hz on a 1:4 scale Perspex model would indicate a resonance of 64 Hz on a full-scale steel frame. A resonance of 150 Hz on a 1:10 scale Perspex model would suggest a resonance of 28·5 Hz on a full-scale cast iron frame.

Perspex models may conveniently be excited by small freely suspended electro-magnetic vibrators, while the motion of the model may be picked up by the use of frequency modulated induction pick-ups with aluminium foil stuck to the surface of the model.

15. Castings versus fabricated frames. For a given weight of metal, frames fabricated from mild steel will be about twice as stiff as cast iron frames merely because of the difference in the elastic moduli. If advantage is taken of welding to reduce the number of joints, they may be made very much stiffer still. If the fabricated design is carried out really intelligently, then the stiffness may be even higher. The damping is likely to be lower, but this penalty is almost certainly worth accepting.

The costs of fabricated frames (or weldments in American terminology) compared with castings are often difficult to estimate at the drawing board stage, since in Britain both castings and weldments tend to be sold by weight if bought out. Typically a weldment may cost 20 percent more than a casting before machining; the fully machined weldment, however, may typically only be 80 percent of the cost of a casting because it can be machined much faster due to its greater stiffness, and many surfaces which would need to be cleaned up on a casting may only need sandblasting on a weldment.

If a price for the weldment can be obtained based not on its weight but on the amount of work in it, which will largely be determined by the inches of weld to be laid down, then the cost can often be greatly reduced by skilful design. Few of the joints will be highly stressed, and stiff joints are required, not strong ones. For this a simple fillet weld is likely to be adequate, saving a lot of edge preparation.

Weldments should be properly stress relieved. If this has been done, and if the welding procedure has been carefully thought out, the distortion of weldments should be no worse than the distortion often met with in castings.

For long runs, money will be well spent on proper welding fixtures which can be offset against the comparable costs of patterns for castings.

Mild steel is not a good material for slideways due to its tendency to scuff, and part of the savings achieved by fabricating may be offset by the necessity to provide cast iron ways on the weldment, unless low friction requirements dictate the use of hydrostatic or rolling bearings.

16. Problems

1. A prototype machine tool has been fabricated out of mild steel. If the production model is to be built in cast iron, how much thicker must the parts of the frame be made to ensure the same static rigidity? How will the resonant frequencies of the machine be affected?

2. The change proposed above will result in core holes being left in the castings. What will be their effects and how can they be assessed?

3. How would you expect the stiffness and natural frequencies of a family of machines to vary with size if the casting wall thickness remains constant regardless of size?

4. Some machine-tool makers deliberately leave the core sand in their machine-tool castings. What effects might this be expected to have on natural frequency and damping?

3. Kinematic trains and drive shafting

1. Machine tools generally have a branched kinematic train. In a lathe, the driving motor drives both the workpiece and also the tool via the lead-screw at a nominally fixed speed relationship to one another. Due to errors in the gears, torsional receptances of the shafting, and other factors, the instantaneous speed relationship will not be exact, and may also be affected by the cutting process. This is the basic mechanism of Arnold-type chatter which will be analysed in Chap. 5. In milling machines, and in particular in hobbing machines, the problem is similar but even more acute because the cut is interrupted, and as a result a roughly sinusoidal forcing function is being applied to both ends of the train.

2. The kinematic train of a production hobbing machine, Fig. 3.1, or hob-type gear grinder, represents the most difficult case that is met with in practice, since adjacent pitch errors, lead errors and involute errors of the order of 0·0002 in (5 μm) are the maximum allowed in gears cut on a good machine. Unless the cutting forces can be kept very low, which implies small cuts and low productivity, the stiffness of all drives must be very high merely in order to avoid variations in the 'wind-up' of the train as the force level varies throughout the cutting cycle. A common trouble in hobbed gears, particularly those with high helix angles, is to find that the helix angle at entry and exit from the gear is different from what it is in the middle due to the variation in cutting force giving rise to variation in the wind-up of the kinematic train.

3. Sources of static wind-up. The obvious sources of wind-up under load are the torsional deflections of the shafts in the kinematic train. These are readily calculable. For geometrically similar machines the stiffness of shafting varies as $\frac{\text{Diameter}^4}{\text{Length}}$, i.e. Scale^3. Since similar accuracies are required from large and small

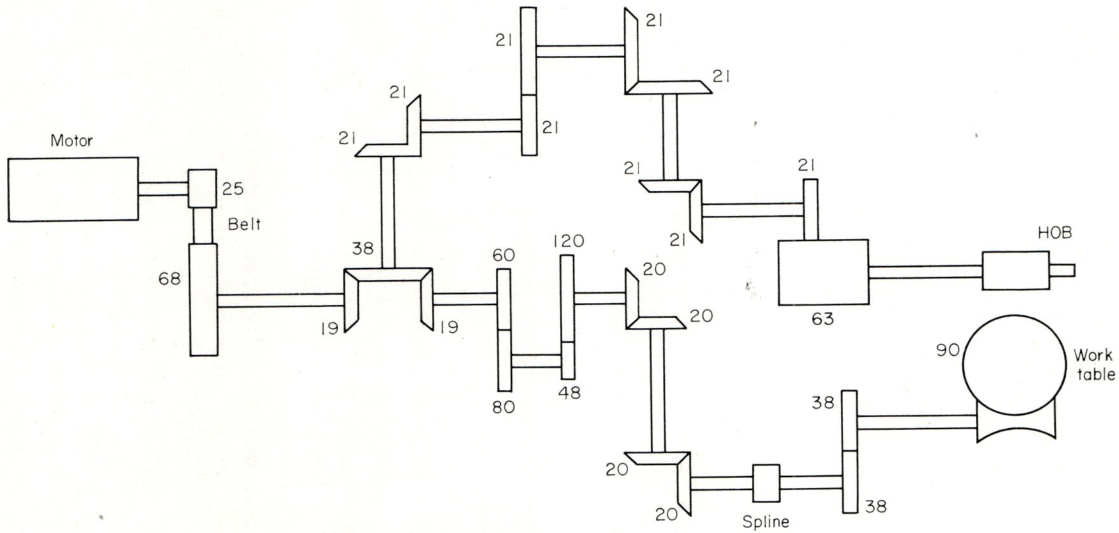

Fig. 3.1. Gearing of hobber kinematic train. The numbers give the number of teeth on each gear.

machines, and since the same sizes of teeth may be cut on both machines, if a machine with a 10 ft table has shafts 6 in diameter, one with a 10 in table will need shafts 2·62 in diameter. This creates difficulties in designing smaller machines to give the same accuracies as larger ones.

The stiffness of the gear teeth in the drives can sometimes be a major source of error; if the safe assumption is made that only one pair of teeth carries the load at any moment, then the rather surprising result appears, from considering a tooth as a simple beam, that the stiffness is independent of the Diametral Pitch.

The stiffness of bearings is also often low, and this can have a major influence, particularly in cases such as the worm drive to the table of a hobber, where the axial position of the worm is determined by the stiffness of a thrust bearing.

4. Torsional vibrations. A first attack on the dynamic behaviour of a kinematic train, such as that shown in Fig. 3.1 can be made using Holzer's method. The assumption will need to be made that no backlash exists in the gears or splines. Unless special precautions have been taken in the design, this will not be true, and under resonant conditions the frequencies may be very different from those calculated, since backlash makes the system nonlinear. It tends to reduce the frequencies calculated, since when two teeth part company neither of them is having a spring force exerted on it, and hence each tends to move more slowly than it would otherwise do. Backlash tends to be more significant in small than in large machines, since the amount tends to remain constant with the size of the machine. Fig. 3.2 shows the rest of the data for an analysis of the train of Fig. 3.1.

From the point of view of accuracy, the train needs to be designed to reduce to a minimum the error in relative angular position between hob and workpiece caused by the cutting forces. No general theory at present exists on how to design a kinematic train to satisfy this condition, but some general guidance on improving the system may be obtained from very simple considerations.

5. The essentials of the problem may be seen by analysing the simple system of Fig. 3.3. If the inertias are I, stiffnesses k, teeth

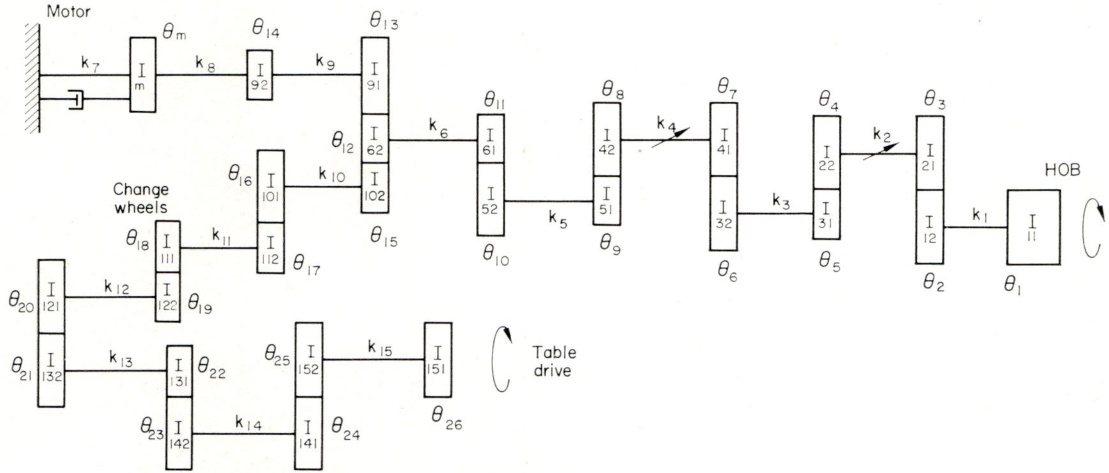

Fig. 3.2. Hobber kinematic train data. θ is angular deflection; k is shaft torsional stiffness; I is moment of inertia of gear, etc. $\not\nearrow$ represents variable stiffness shaft (spline coupled shaft).

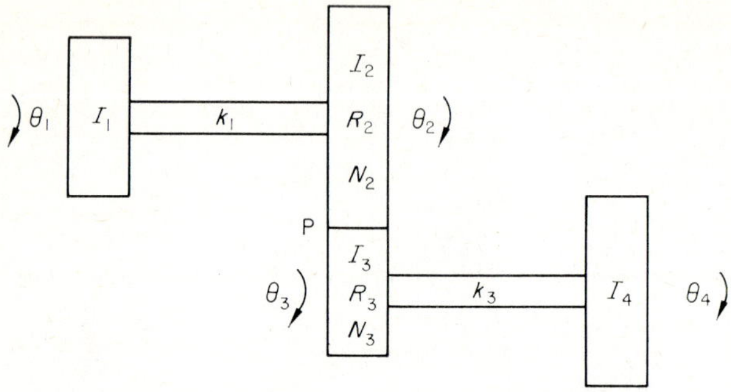

Fig. 3.3 Effect of gearing on equivalent inertias, etc.

numbers N, the pitch circle radii R, all with appropriate suffixes,

$$I_1 D^2 \theta_1 = -k_1(\theta_1 - \theta_2) \tag{3.1}$$

If P is the force acting at the tooth contact,

$$I_2 D^2 \theta_2 = +k_1(\theta_1 - \theta_2) - PR_2 \tag{3.2}$$

$$I_3 D^2 \theta_3 = -k_3(\theta_3 - \theta_4) - PR_3 \tag{3.3}$$

$$I_4 D^2 \theta_4 = +k_3(\theta_3 - \theta_4) \tag{3.4}$$

Adding Eq. 3.3 and 3.4 gives

$$D^2(I_3\theta_3 + I_4\theta_4) = -PR_3$$

$$= -P\frac{N_3}{N_2}R_2 \tag{3.5}$$

Adding Eq. 3.1 and 3.2 gives

$$D^2(I_1\theta_1 + I_2\theta_2) = -PR_2 \tag{3.6}$$

∴ from Eq. 3.5 and 3.6

$$D^2\left(I_1\theta_1 + I_2\theta_2 - \frac{N_2}{N_3}\,.\,I_3\theta_3 - \frac{N_2}{N_3}\,.\,I_4\theta_4\right) = 0$$

But $\theta_3 = -\frac{N_2}{N_3}\,.\,\theta_2$

and from Eq. 3.4

$$\left[\frac{I_4}{k_3}\,.\,D^2 + 1\right]\theta_4 = \theta_3 = -\frac{N_2}{N_3}\,.\,\theta_2,$$

$$\therefore D^2\left\{I_1\theta_1 + I_2\theta_2 + \frac{N_2^2}{N_3^2}\,.\,I_3\theta_2 + \frac{N_2^2}{N_3^2}I_4\frac{\theta_2}{\left[\frac{I_4}{k_3}\,.\,D^2+1\right]}\right\} = 0 \tag{3.7}$$

Inspection of this equation suggests that the inertias may be *normalized* by multiplying them by the ratio of their (mean speed)² to the (mean speed of driving shaft)^2. The last term of this equation shows that stiffnesses may also be similarly normalized so that $\frac{I_4}{k_3}$ may be written

$$\frac{\left(\frac{N_2}{N_3}\right)^2 I_4}{\left(\frac{N_2}{N_3}\right)^2 k_3}$$

Normalizing the inertias gives directly the relative flywheel, or energy storage, effect of the various components of the system.

6. Much information may be gained about the possibilities of improving the design of a kinematic train by means of a diagram such as is shown in Fig. 3.4, devised by Mr J. E. Sharpe. In it is represented the data of the train of Figs 3.1 and 3.2. The top part of the diagram represents the main kinematic train from hob to workpiece. Vertically are shown the normalized inertias of the various wheels in the train into which are also lumped the inertias of the shafts attached to them. The distances separating these inertias along the horizontal axis are proportional to the inverse of the normalized stiffnesses (i.e. the normalized flexibilities). A large horizontal spacing between two inertias represents relatively large flexibility in the shafts connecting them. This flexibility comes partly from the moment of inertia of the shaft cross section, partly from its length.

The bottom portion of the diagram, connected to the top half by a dotted line, represents schematically the motor branch of the kinematic train.

Inspection of Fig. 3.4 shows that for a first analysis the top branch of the system might well be represented approximately by five inertias and four shafts. The very large inertia in the middle comes somewhat surprisingly from the change gears of the machine; this is therefore of variable amount, but an investigation of the range of possibilities at this point shows that this will always be the largest effective inertia in this particular design, and that as a result the system may reasonably be split into two sub-systems, one on either side of this point.

The designer will always try to make the resonant frequencies of a machine as high as possible; he will also try to obtain nodes in

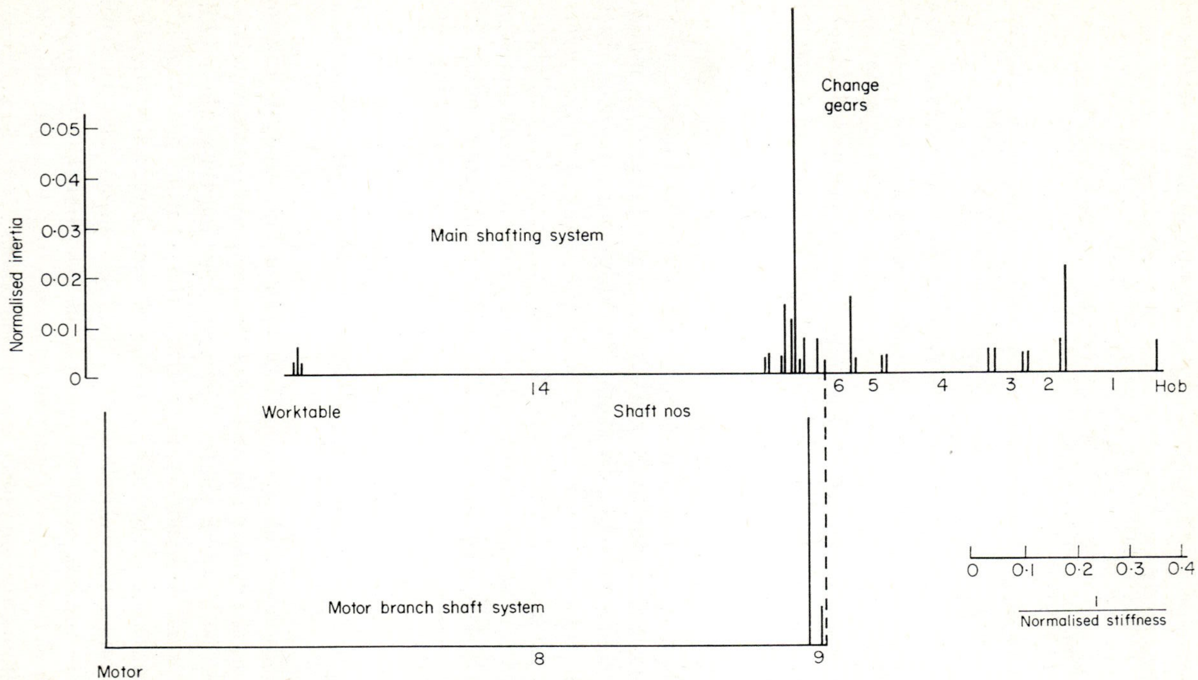

Fig. 3.4. Hobber relative inertias and compliances.

the vibration pattern both at the table and also at the hob, since at a node the amplitude of the vibration is theoretically zero. The diagram makes clear that the most profitable points at which to attack the problem of increasing the resonant frequencies is likely to come from increasing the stiffnesses of shafts 1, 4 and 14. From the point of view of trying to force a node at the hob, a flywheel on the hobshaft, or if this is not possible on the next shaft back from it, will probably be useful. A vibration damper occupying the same amount of space might be more helpful, but its effect cannot be assessed from this sort of diagram; and indeed is very difficult to assess at all; this problem is discussed in Chap. 8.

7. Effect of drive motor. Careful measurements on a large hobbing machine fitted with a variable speed d.c. motor showed that very large errors in position between hob and table, of the order of 10 μm, were introduced by an impulsive voltage variation of only 1 V in the 440 V d.c. supply due to the switching-on of another machine. In a normal factory supply, unless special precautions are taken, the fluctuations in mains voltage are frequently much larger than this. The reason for these errors is the fact that very small step voltage changes on some types of motor produce very large transient torque changes (Para. 9). These are applied at the middle of a system in which the two branches normally have very different stiffnesses and inertias with the result that very much larger changes will be introduced in one part of the system than in the other, and the changes will not be in phase.

8. Transient characteristics of electric motors. The commonest form of machine-tool driving motor is an *induction motor* either with or without slip rings; its dynamic characteristics sometimes play a significant part in the complete dynamics of the machine.

Small changes of supply voltage, such as are likely to be occurring continually in a normal workshop due to variations of demand on the mains, give rise to small changes of torque with a time constant of about $\frac{1}{2}$ s.

Since Torque $\propto V^2 \times$ Slip,

$$\frac{t}{T} = \frac{2v}{V}$$

where T = mean torque, t = small change in torque, V = mean voltage, v = small change in voltage.

During the transient period while the speed is changing, various current and torque changes, mainly oscillatory at mains frequency, will occur. These are usually well damped with a period of about $\frac{1}{2}$ s.

Speed changes will also occur due to torque variations: a complete analysis of these shows a transfer function with a cubic equation on the bottom line, and a second order equation on the top: for most practical purposes a good approximation is given by the transfer function below so long as slip $\not> 0{\cdot}05$ at maximum torque. If

T = torque
t = small change in torque
L = rotor leakage inductance
R = rotor resistance
J = inertia
Ω = supply frequency
ω = small change in motor speed in rad/s
k = slope of steady state torque/slip curve at slip = 0

then

$$\frac{\dfrac{\omega}{\Omega}}{\dfrac{t}{J\Omega^2}} = -\frac{\left[\dfrac{\Omega L}{R}\mathrm{D}+1\right]}{\mathrm{D}\left[\dfrac{\Omega L}{R}\mathrm{D}+1\right]+\dfrac{k}{J\Omega^2}}$$

9. When variable speed drives are required, with fine speed regulation, *direct current shunt motors* are sometimes used. These motors are very sensitive to small variations in the supply voltage. The field has a large inductance and due to saturation does not change very much for small voltage changes. Its time constant is of the order of 0·5 s and the percentage change in field will be about half the percentage change in voltage. The armature has a low inductance, with a time constant of about 0·05 s, and the voltage drop across it is mostly absorbed by the back EMF: for a 5 HP motor, only 5 percent of the armature voltage drop will be due to the resistance of the rotor. A sudden drop in voltage of 1 percent will therefore cause a sudden drop in torque of about 20 percent with a time constant of 0·05 s. As the field strength falls, the back EMF will drop and the torque will increase; the falling speed of the system will further decrease the back EMF.

For a motor alone, the mechanical time constant for this latter process would be about 0·1 s, but the inertia of the driven system will usually increase this.

10. Synchronous a.c. machines are occasionally used because of their value in power factor correction. Their percentage change in torque is only likely to be about 1/10th of the percentage change in voltage, and the variation of slip angle would be very small. Transient effects would normally die away with a time constant of about 1/5th s.

11. Problems

1. Hydrostatic drives are being advocated for machine tools. Using manufacturer's literature, compare them with electric motors from the point of view of (*a*) size, (*b*) moment of inertia, (*c*) mean torque produced, (*d*) torque fluctuations and possible exciting frequencies, (*e*) efficiency, including supplies, (*f*) possibilities of fitting so as to avoid wind-up of kinematic train, (*g*) time constants and transfer operators.

2. If torsional vibrations are occurring in a drive shaft system at frequencies which are always proportional to shaft speed (e.g. once per rev. or 12 times per rev.) why is the centrifugal pendulum type of vibration absorber theoretically better than the normal types? (N.B. For the theory of the pendulum type of damper see den Hartog *Mechanical Vibrations*.)

4. Dynamics of the cutting process

1. Most of the early work on the theory of metal cutting hinged round the concept of a steady process in which a chip of metal is produced by shearing a strip of uncut metal continuously and uniformly, the deformation taking place along a shear plane. This is the essence of Merchant's Theory, discussed below, which in many cases is a good approximation to what happens when cutting metal in a lathe, although this is not a strictly two-dimensional case.

In Fig. 4.1 the uncut metal is of thickness d; as it comes up to the tool it is sheared, and leaves parallel to the face of the tool with thickness t. The width of the chip is supposed to be a constant b throughout the process, and neither face of the metal being cut is supported. The face of the tool is perpendicular to the plane of cutting. These assumptions give so called orthogonal cutting.

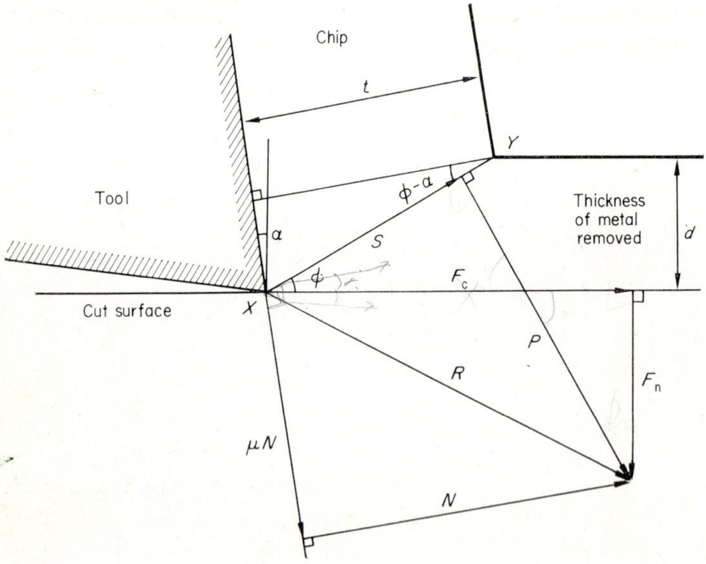

Fig. 4.1. Merchant's Theory. XY is shear plane.

The obvious criticism of them is that metal cannot shear instantaneously, and that the shearing must take place in a finite time within a shear zone. Elementary observation notes that chips normally curve away from the surface of the tool. Detailed observation shows that the metal often sticks to the tip of the tool and forms what is called a 'built-up edge' which forms and breaks away again, the process being repeated in a fairly regular manner.

Many more refined theories have been propounded, but all appear to be open to the objection that they do not take account of the random variations which occur in most cutting processes, nor of the fact that no machine or experimental rig can be rigid enough to verify them.

2. Merchant's Theory. Making the assumptions of the previous paragraph, analytical relationships may be developed between the components of cutting force, the coefficient of friction, μ, the shear strength of the material, τ, and the geometry of the system. Taking the notation of Fig. 4.1, if F_c and F_n are the components of cutting force in the direction of and at right angles to the movement of the tool, which may be measured experimentally,

$$S = F_c \cos\phi - F_n \sin\phi$$
$$P = F_c \sin\phi + F_n \cos\phi$$
$$N = F_c \cos\alpha - F_n \sin\alpha$$

If μ is the coefficient of friction at the tool,

$$\mu N = F_c \sin\alpha + F_n \cos\alpha$$

$$\therefore \mu = \frac{F_c \sin\alpha + F_n \cos\alpha}{F_c \cos\alpha - F_n \sin\alpha}$$

$$= \frac{F_c \tan\alpha + F_n}{F_c - F_n \tan\alpha} \qquad (4.1)$$

From geometry,

$$XY = \frac{d}{\sin\phi} = \frac{t}{\cos(\phi-\alpha)}$$

$$\frac{d}{t} = r = \frac{\sin\phi}{\cos(\phi-\alpha)}$$

and may be obtained experimentally.

$$\tan\phi = \frac{\cos\alpha}{1 - r\sin\alpha} \qquad (4.2)$$

$$\text{Area of Shear Plane} = b \,.\, XY = \frac{bd}{\sin\phi} = A_s$$

$$\therefore \tau = \frac{s}{A_s} \qquad (4.3)$$

and

$$\sigma = \frac{P}{A_s} \qquad (4.4)$$

If the Cutting Velocity $= V$, then the Chip Flow Velocity $= rV$. Note that for mild steel, $\phi \doteqdot 22\frac{1}{2}°$. Its range is roughly 15°–30°.

3. Cutting forces. Despite very large variations of the instantaneous values of cutting force, mean values can be measured accurately. Fig. 4.2 defines the three components of cutting force when hobbing and Fig. 4.3 shows how the three components vary as the cutting speed changes.

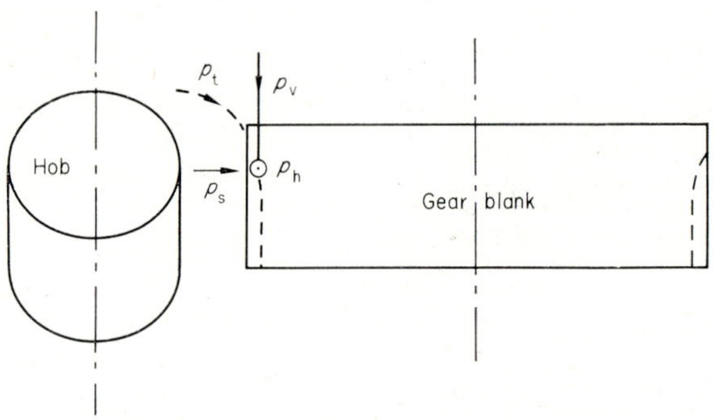

Fig. 4.2. Measured hobbing force orientations.

Fig. 4.4 shows the variation of the components of cutting force with the depth of cut, which is proportional to axial feed. The curves rise steadily, but do not go through zero. This may be explained by the fact that the edge of the cutter could be deforming the surface of the workpiece elastically, plastically, or both, without actually cutting it, as in normal rubbing.

While the mean and peak values obtained when cutting steadily may be very consistent, the detailed moment to moment variation of cutting forces may be very complex indeed, as is shown in Fig. 4.5. The curves shown above were extracted from force

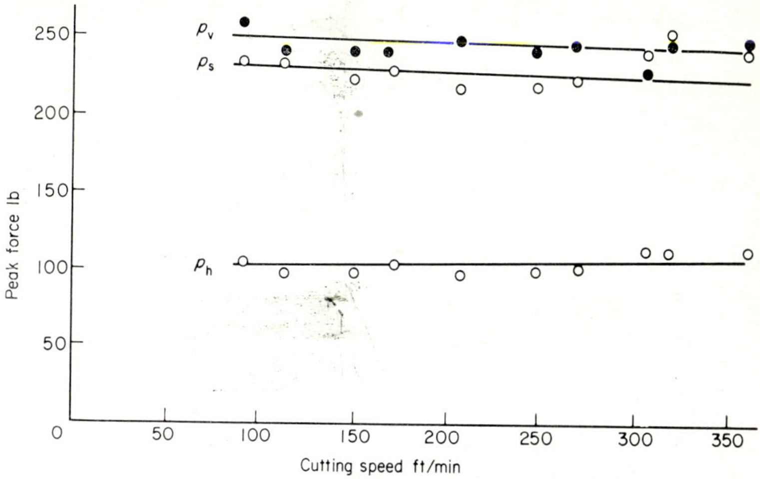

Fig. 4.3. Peak cutting forces *v*. cutting speed. Climb hobbing. Machining data: 8 D.P. Spur, E.N. 8; Hob 3 in O.D. 12 Gash, Single Start; Feed 0·052 in/rev.

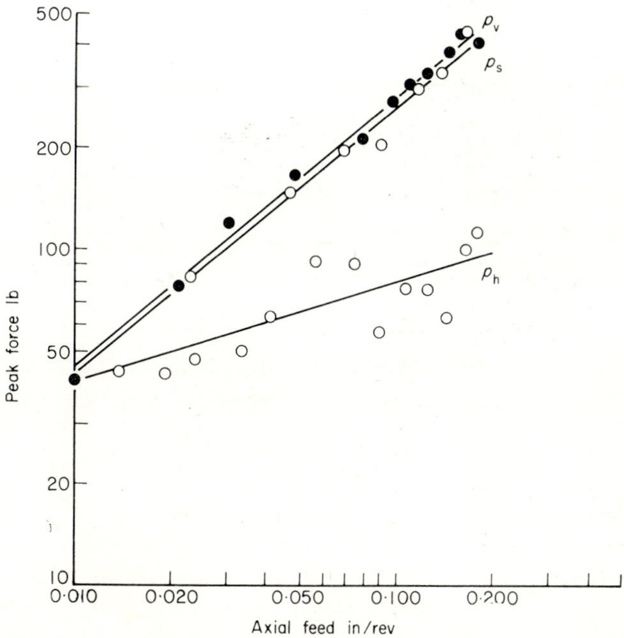

Fig. 4.4 Peak cutting forces *v*. axial feed. Climb hobbing. Machining data: 12 D.P. Spur, E.N. 8; Hob 12 Gash. $2\frac{3}{4}$ in O.D., Single Start; 114 r.p.m. N.B., logarithmic scales.

measurements of this nature. Except at high speeds, all measurements of forces during so-called steady cutting show very considerable amounts of 'noise'. Any theory of cutting which does not account for this must be considered to be very dubious.

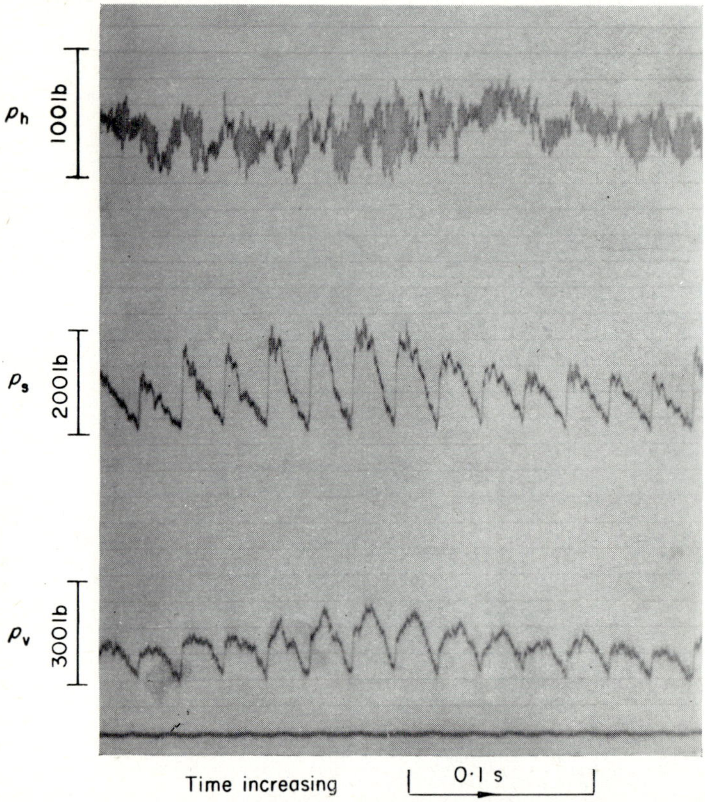

Fig. 4.5. Cutting forces *v.* time. Climb hobbing. Machining data: 0·050 in/rev. axial feed; Hob 8. D.P., Single Start, 12 Gash, 3 in O.D., 220 r.p.m.; Gear Spur, E.N. 8, whole depth of tooth 0·270 in.

4. Dynamic cutting forces. When a machine tool is chattering steadily, the surface generated is roughly sinusoidal in nature. Much study has therefore been put into the forces which occur, Fig. 4.6 (i), when the tool is cutting a nominally uniform smooth surface beneath a wavy surface; and, Fig. 4.6 (ii), when the tip of the cutting tool is moving on a sinusoidal path beneath a nominally smooth surface. In both cases measurements show that

a. the force traces obtained may be very far from sinusoidal, and very 'noisy';
b. a very heavy second harmonic component of force may be present;
c. the force components are not necessarily in phase with the sinusoidal variation of surface or of tool;
d. the phase shift may sometimes be leading and sometimes lagging.

Various proposals have been put forward to account for these facts. The very simplified analyses which follow indicate how useful qualitative ideas may be obtained from simple assumptions.

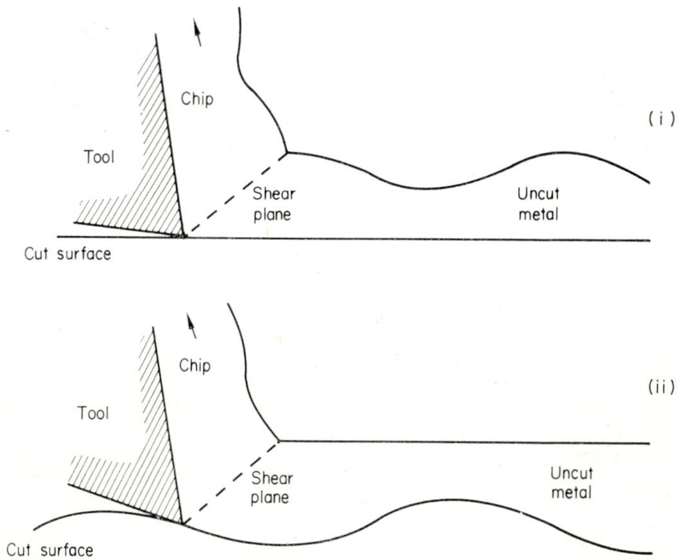

Fig. 4.6. Chip thickness variations.

5. Cutting forces when surface is varying sinusoidally. This is the case of Fig. 4.6 (i); the analysis is shown in Fig. 4.7. The approximate assumptions are made that the tip of the tool is moving at constant velocity, and that the metal is sheared on a plane which remains at a constant angle. As has been noted, shear takes place in a zone, and observation shows that this tends to wander under these conditions. Furthermore, make the assumption that the main component of cutting force in the direction of the cut is proportional to the length of the shear line: this may also be expressed as the power being proportional to the rate of metal removed. Accepting these very approximate assumptions, the diagram shows that as the shear plane moves the variation in its length is $2a$ cosec θ. This does not happen sinusoidally, since the rate of change is much higher when the slope of the shear plane and of the surface are of the same sign than when they are of different sign. Since the process repeats itself regularly, the fundamental term of the Fourier series representing what happens can be deduced directly from the geometry. Because of the change in relative slopes, a very strong second harmonic may be expected, and measurements show that this is indeed often the case. A phase difference exists between the height of the surface above the tip of the cutting tool, and the point at which the shear plane meets the surface. From the figure the fundamental component of the variation in shear plane length can be seen to be

$$(a \operatorname{cosec} \theta) \sin \left(\frac{2\pi V}{\lambda} t + \frac{2\pi A \cot \theta}{\lambda}\right)$$

The force variation is assumed to be proportional to this.

While this expression has been deduced for a sinusoidal surface, it may be generalized for any form of surface in the manner usual for a finite time delay, noting that in this case it usually represents a time lead. This gives the transfer function for variation in shear plane length as

$$(a \operatorname{cosec} \theta)\, e^{\frac{-2\pi A \cot \theta}{\lambda} \cdot D}$$

(see *ECT** p. 49).

* See Bibliography.

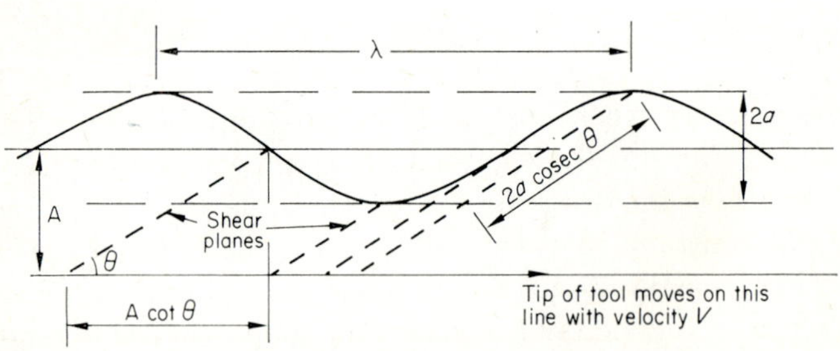

Fig. 4.7. Shear plane length variation. Fundamental component of variation in shear plane length $= (a \operatorname{cosec} \theta) \sin \left(\frac{2\pi V}{\lambda} t + \frac{2\pi A \cot \theta}{\lambda}\right)$

Both the component of force parallel to the surface and also that perpendicular to the surface may be expected to vary in this manner. Whether the phase is leading or lagging will depend on the depth of cut and wavelength, as well as on the amplitude of the surface wave. Some investigators maintain that it also depends on the instantaneous slope of the surface wave.

6. Cutting forces when tool is moving sinusoidally in feed direction. When the tool is removing a chip as in Fig. 4.6 (ii), both the main component of cutting force parallel to the metal surface and the perpendicular component vary approximately sinusoidally. The phase of the forces relative to the displacement can be either lagging or leading. Since the cutting process in steel is extremely complex, this behaviour is not surprising.

Many theories have been put forward to account for observed experimental results but none is entirely satisfactory. The results measured can be explained qualitatively with steel as follows:

a. At low cutting speeds and low frequencies. Under these conditions the built-up edge takes time to alter its size and shape as the chip thickness varies. When the chip thickness is decreasing, the built up edge is larger than when the thickness is increasing, so that the effective rake angle is larger and hence the cutting forces are smaller. This effect gives a force which is lagging on the displacement; this is normally destabilizing.

b. At low cutting speeds and high frequencies. Under these conditions the built-up edge cannot change fast, so does not affect the phase. The 'friction' between chip and tool rake face occurs at high temperatures with the metal in a semi-plastic state. As a result, the measured 'friction' has viscous characteristics, since the metal behaves like a thick grease and the force is proportional to the relative velocity between tool and chip. This, like a normal dashpot damper, gives a force leading the displacement, and usually has a stabilizing effect on vibrations.

c. At high cutting speeds. The built-up edge disappears at high cutting speeds, as is shown by the chip coming off straight instead of curled; as a result the mechanism described in (*a*) no longer exists. At high cutting speeds, the relative velocity of tool and chip due to vibration is very small compared with the mean velocity of the chip, so that the effects described in (*b*) are small.

The observed phase changes can be explained by considering surface temperature variations on the tool rake face, as the contact between tool and chip varies with chip thickness. When the chip thickness is decreasing, the tool rake face, having been in contact with the chip, is hotter than when the chip thickness is increasing and the chip is encountering a relatively cold tool. With the higher temperature the plastically flowing contact layers between chip and rake face are weaker and so give less 'friction'. There is thus less friction force when the chip thickness is decreasing, i.e. the force leads the displacement, giving extra damping.

For analysis purposes, if the variation in the component of force perpendicular to the workpiece is p_s, the cutting force stiffness (which is assumed to be constant) is k_s and the variation in depth of cut is z

$$p_s = k_s \, e^{\pm j\phi} \, z$$

where ϕ is the lead or lag angle, which must usually be obtained experimentally.

7. Interrupted cutting in milling and hobbing. The traces of hobbing shown in Fig. 4.5 show curves for which the force during the individual cut is very far from sinusoidal, and for which the content of higher harmonics is high. In addition a lot of noise is present. These processes therefore introduce complex forcing terms on the workpiece and machine, which can excite resonances.

8. Problem

1. The curves of Fig. 4.4 do not extrapolate to zero force at zero feed. Why is this? Under what conditions will a cutting tool: (*a*) remove metal, (*b*) deform the surface elastically, (*c*) cause plastic deformation of the surface? Apply this thinking to the problem of shaving gears; how does a shaving cutter cut?

5. Chatter

1. Three forms of chatter are discussed in this chapter, all of which may be analysed by simple linear theory; quite possibly other non-linear forms also exist. The first is Arnold-type chatter, named after the man who discovered it, which is due to the variation of force with cutting speed. The second type is usually known as regenerative chatter, but is sometimes associated with the names of Tobias and Tlusty who described it independently; regenerative chatter occurs when the cutting force varies as a result of unevenness of the surface being cut due to variations in the cutting force on the last occasion on which the tool passed over that particular part of the surface. The third type of chatter is due to mode coupling when forces acting in one direction on a machine-tool structure cause movements in another direction, and vice versa.

All these forms of chatter are due to feedback in the machine tool between the cutting process and the tool frame and drive considered as a closed loop system as shown in Fig. 1.1.

Machine tools vary so much not only between different types but also between two successive tools off the same production line, that few figures are quoted in this chapter. Its aim is to provide a qualitative idea of the essential nature of the problems which are met. One useful order of magnitude figure is that the static stiffness of most machine tools measured between the cutting tool and the workpiece tends to be around 10^5 lbf/in. A figure of 10^6 lbf/in is exceptionally good, one of 10^4 lbf/in is rather bad, but may be excusable in certain types of small machine tools.

2. Velocity dependent chatter in lathes (Arnold-type chatter). Making initially the most elementary possible assumptions about the system shown in Fig. 5.1, the work piece and its drive system are regarded as being rigid and rotating at constant angular velocity. The tip of the tool can move as shown in the plane of the paper, but is completely rigid in all other planes. As has been shown in Chap. 2, the tool and its support may be regarded as a simple spring-mass system with damping so long as its vibration is occurring near to its resonant angular frequency ω_x. Taking the effective mass of the system as M_x, the effective damping coefficient as λ_x, and the stiffness as k_x, the equation of motion may be written down in the usual manner. The cutting force at any instant is $P+p$, where P is the mean steady-state force, and p is the instantaneous variation from it.

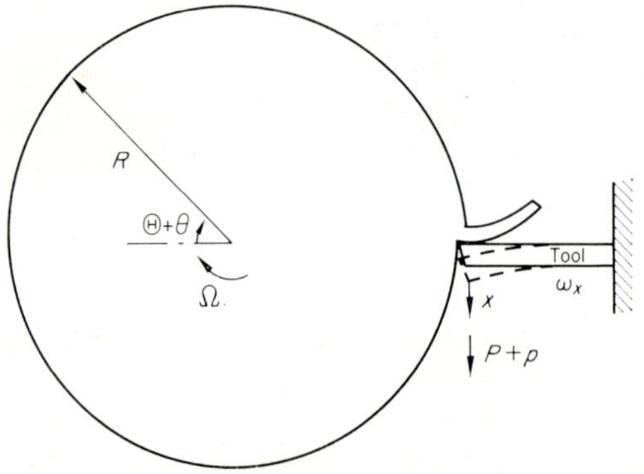

Fig. 5.1. Arnold-type chatter.

$$[M_x D^2 + \lambda_x D + k_x]x = p \tag{5.1}$$

The tool moves with velocity Dx, and hence the relative cutting velocity drops by Dx. This means, Fig. 5.2, that the cutting force increases: so if λ_a is a constant relating cutting force and speed,

$$p = \lambda_a Dx \tag{5.2}$$

Eliminating p from these two equations gives

$$[M_x D^2 + (\lambda_x - \lambda_a)D + k_x]x = 0 \tag{5.3}$$

This represents unstable vibratory motion if

$$(\lambda_x - \lambda_a) \leqslant 0 \tag{5.4}$$

For *stability*,

$$\lambda_x > \lambda_a \tag{5.5}$$

Increasing the damping of the machine will therefore increase its stability.

Chatter is normally a steady condition, whereas Eq. 5.3, when the damping term is negative, represents steadily increasing motion. In practice the motion becomes non-linear, and the non-linearity determines the amplitude of the motion.

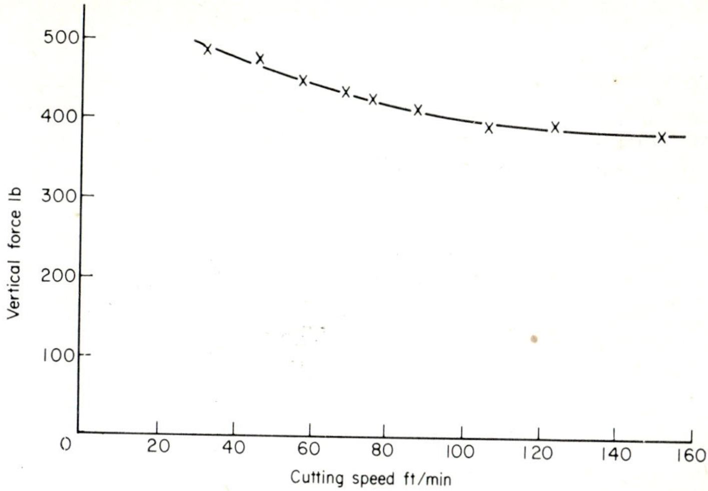

Fig. 5.2. Variation of shaping force with speed. Machining data: E.N. 8, 235 V.P.N.; 29T Spur 8 D.P.; First cut 70 percent depth; $7\frac{1}{4}$ thou. radial feed.

3. This analysis may readily be extended to take note of the variations in angular velocity of the workpiece due to the flexibility of the drive system. If at a given moment the position of the workpiece is given by $\Theta+\theta$, the instantaneous angular velocity will be $\Omega+\mathrm{D}\theta$. Eq. 5.2 therefore modifies to

$$p = \lambda_a(\mathrm{D}x - R\,.\,\mathrm{D}\theta) \tag{5.6}$$

If the drive system has an undamped natural frequency ω_θ, damping ratio c_θ, and spring stiffness k_θ, then

$$\left[\frac{\mathrm{D}^2}{\omega_\theta^2}+\frac{2c_\theta}{\omega_\theta}+1\right]\theta = -\frac{1}{k_\theta}[R\,.\,p+k_e\mathrm{D}\theta] \tag{5.7}$$

where k_e is an appropriate transfer operator for the electric motor and belt drive (see Chap. 3, Para. 8).

Rewriting Eq. 5.1 in the standard form, where ω_x is the undamped natural frequency, c_x is the damping ratio, gives

$$\left[\frac{\mathrm{D}^2}{\omega_x^2}+\frac{2c_x}{\omega_x}\mathrm{D}+1\right]x = \frac{1}{k_x}\cdot p \tag{5.8}$$

Eliminating p and θ from these equations gives a fourth order differential equation representing a motion for x which may be stable or unstable depending

a. on whether coefficients are positive or negative; and

b. on the Routh criterion for the ratios of coefficients.

4. Regenerative chatter in lathes. In Fig. 5.3 the tool is being fed to the left at a steady speed. It is considered as having a natural angular frequency ω_y in the y direction, but otherwise to be completely rigid. At any moment, due to the cutting action, a mean force F and an instantaneous force $(F+f)$ is acting in the y direction. The assumption is made that the cutting force variation f is proportional to the instantaneous variation in the thickness of the chip being cut. This thickness variation will depend on the present position of the tool, given by y, *but also on the position of the tool one revolution previously*. The workpiece is considered to have constant angular velocity Ω. The force on the tool and its displacement are related by the equation

$$f = \underbrace{-k_t y}_{\substack{\text{due to}\\\text{present}\\\text{position}}} + \underbrace{k_t e^{\frac{-2\pi}{\Omega}\mathrm{D}} y}_{\substack{\text{due to position}\\\text{one revolution}\\\text{earlier.}}} \tag{5.9}$$

where k_t is a constant called the *chip thickness factor*, or *cutting force factor*. It has the dimensions of a stiffness.

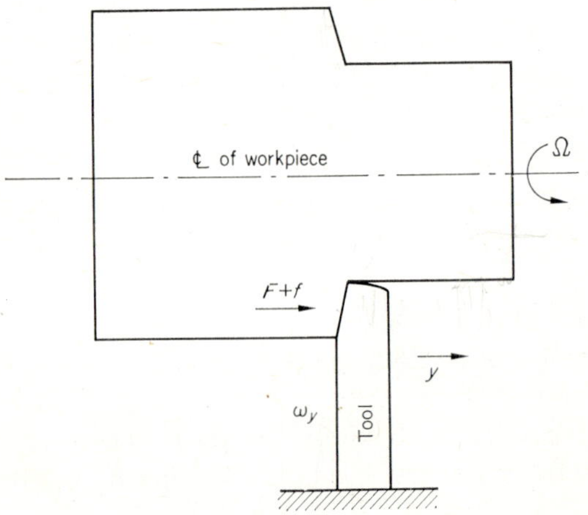

Fig. 5.3. Lathe regenerative chatter.

The equation of motion of the tool is

$$\left[\frac{\mathrm{D}^2}{\omega_y^2} + \frac{2c_y}{\omega_y}\mathrm{D}+1\right]y = \frac{1}{k_y}\cdot f \tag{5.10}$$

where k_y is the static stiffness of the system, and c_y the effective damping ratio, which contains the damping due *both* to the frame, c_f, *and* to the cutting process, c_c, i.e. $c_y = (c_f + c_c)$.

Eliminating f gives

$$e^{\frac{2\pi}{\Omega}D}\left[\frac{D^2}{\omega_y^2}+\frac{2c_y}{\omega_y}D+\left(1+\frac{k_t}{k_y}\right)\right]y-\frac{k_t}{k_y}y = 0 \tag{5.11}$$

This equation has a very simple physical meaning, as can be seen by reference to Fig. 5.4, when the loci are drawn for steady sinusoidal motion; this can only occur when the intercept along the real axis at which the vector locus crosses it is equal to $\frac{k_t}{k_y}$. Now at resonance of the machine, the terms in the square bracket must contribute about $\frac{\pi}{2}$ shift in phase angle. The finite time delay must therefore swing the vector through about $\frac{3}{2}\pi + 2\pi n$, where $n = 0,1,2,\ldots$ (which it does without change of amplitude), for chatter to be possible: when $n = 0$, $\frac{2\pi}{\Omega}\omega = \frac{3\pi}{2}$. The real intercept therefore is now $\frac{2c_y\omega}{\omega_y}$.

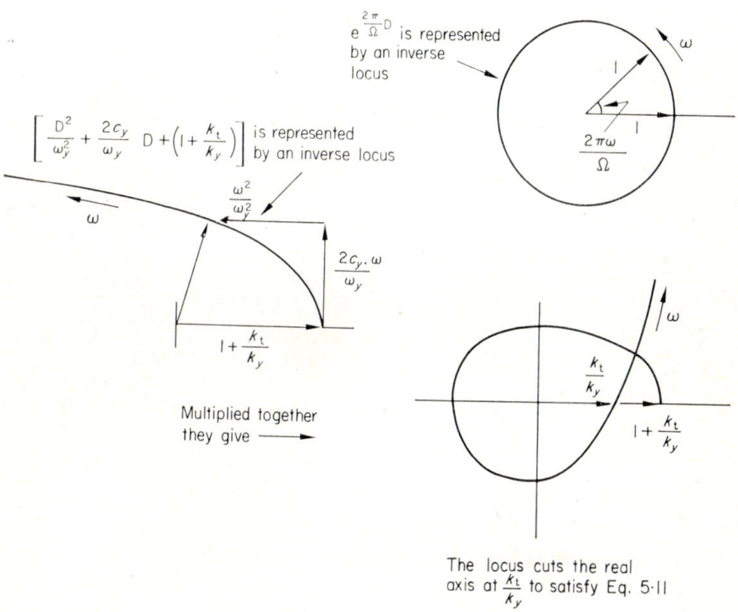

Fig. 5.4. Chatter inverse loci.

[N.B. Another way of visualizing what occurs is to imagine that an external force $e^{\frac{2\pi}{\Omega}D}\,.\,P\,e^{j\omega t}$ is applied to the tool. This causes Eq. 5.10 to be modified to

$$\left[\frac{D^2}{\omega_y^2}+\frac{2c_y}{\omega_y}D+1\right]y=\frac{1}{k_y}\cdot f+\frac{1}{k_y}\cdot e^{\frac{2\pi}{\Omega}D}\,.\,P\,e^{j\omega t}$$

and Eq. 5.11 to be modified to

$$e^{\frac{2\pi}{\Omega}D}\left[\frac{D^2}{\omega_y^2}+\frac{2c_y}{\omega_y}D+\left(1+\frac{k_t}{k_y}\right)\right]y-\frac{k_t}{k_y}\cdot y=\frac{P}{k_y}\cdot e^{j\omega t}$$

The condition for self-sustained sinusoidal motion is that $P=0$].

For steady sinusoidal motion, to satisfy the equation,

$$\frac{2c_y\omega}{\omega_y}=\frac{k_t}{k_y}$$

$$\therefore\quad \frac{\omega}{\omega_y}=\frac{1}{2c_y}\cdot\frac{k_t}{k_y} \tag{5.12}$$

Also from Fig. 5.4, for this condition

$$\frac{\omega^2}{\omega_y^2}=\left(1+\frac{k_t}{k_y}\right)$$

$$\therefore\ \omega=\sqrt{1+\frac{k_t}{k_y}}\,\omega_y \tag{5.13}$$

Experimentally k_y is found to be of the order of 2×10^5 lbf/in. $\frac{k_t}{k_y}$ is found to be of the order of $\frac{1}{2}$ to $\frac{1}{50}$, and $\frac{1}{2c_y}$ to be about 2–50 as is necessary to satisfy Eq. 5.12 and Eq. 5.13 simultaneously. $\left(\text{The `}Q\text{ factor' of the system is given by } Q=\frac{1}{2c_y}.\right)$

$$\text{Now}\ \frac{2\pi}{\Omega}\,\omega=\frac{3\pi}{2}$$

$$\therefore\ \omega=\tfrac{3}{4}\Omega \tag{5.14}$$

This means, as is found in practice, that the sinusoidal irregularity of the surface of the workpiece moves circumferentially relative to the workpiece.

SHEFFIELD UNIV.
APPLIED SCIENCE
LIBRARY.

5. Chatter can occur over a range of speed, as is shown in the RH shaded zone of Fig. 5.5 due to the tool contributing slightly less or more than $\frac{\pi}{2}$ phase shift, and the finite time delay being correspondingly greater or smaller; this is shown in Fig. 5.6. Since on either side of resonance the input vector is longer than at resonance, when the vector is swung round to lie on the real axis, the value of $\frac{k_t}{k_y}$ to satisfy the stability equation is greater.

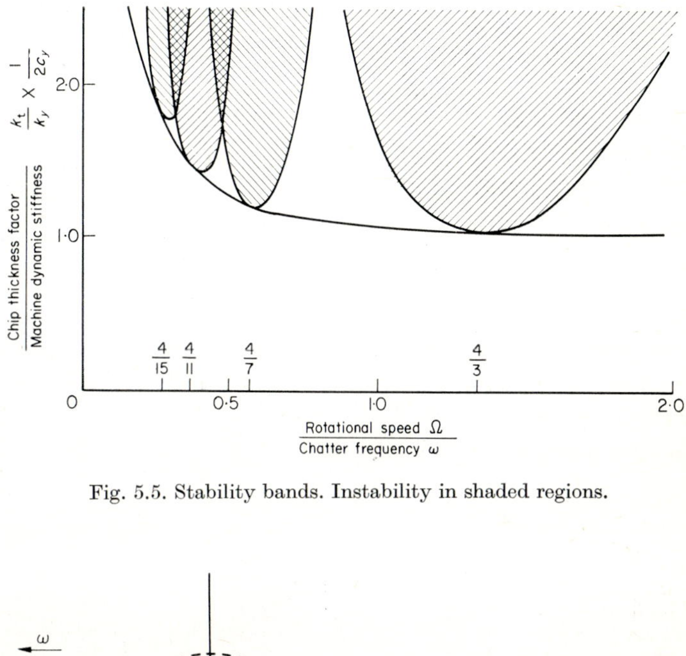

Fig. 5.5. Stability bands. Instability in shaded regions.

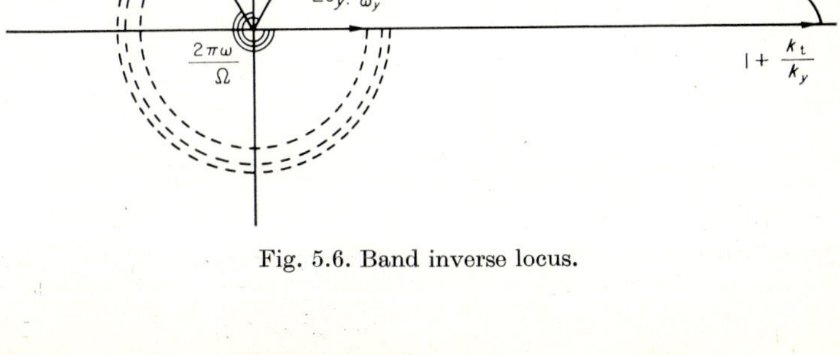

Fig. 5.6. Band inverse locus.

The other areas of instability in Fig. 5.5 are due to the finite time delay being given by $\frac{3\pi}{2}+2\pi n$. This gives expectation of chatter for

$$\left.\begin{aligned} n &= 0, \frac{\Omega}{\omega} = \frac{4}{3} = 1{\cdot}33 \\ n &= 1, \frac{\Omega}{\omega} = \frac{4}{7} = 0{\cdot}57 \\ n &= 2, \frac{\Omega}{\omega} = \frac{4}{11} = 0{\cdot}36 \\ n &= 3, \frac{\Omega}{\omega} = \frac{4}{15} = 0{\cdot}27 \text{ etc.} \end{aligned}\right\} \tag{5.15}$$

The reason for the minimum values of the lobes not being identical is that k_t, the cutting stiffness, and c_c, the component of c_y due to the cutting process, vary with the mean cutting speed. The contribution of the cutting process to the damping varies approximately inversely with the cutting speed.

6. For chatter *not* to occur, from the vector locus

$$\frac{2c_y\omega}{\omega_y} > \frac{k_t}{k_y} \tag{5.16}$$

This leads to a useful and simple guide for the machine-tool designer.

We may write Eq. 5.10 in the form

$$[M_y D^2+\lambda_y D+k_y]y = f \tag{5.17}$$

where M_y is the effective mass and λ_y the damping coefficient, given by

$$\lambda_y = 2c_y\frac{k_y}{\omega_y} \tag{5.18}$$

and substituting in Eq. 5.16 gives $\omega \simeq \omega_y$,

so

$$2c_y\omega = \frac{\lambda_y\omega}{k_y}$$

and

$$\lambda_y\omega_y > k_t \tag{5.19}$$

In words,

Machine Resonant Dynamic Stiffness > Cutting Force Factor.

Increasing the damping of the machine will therefore reduce the tendency of a machine to chatter, whether the chatter be Arnold or regenerative type. Increasing the stiffness reduces the chance of regenerative type chatter, but also often reduces the damping, due to elimination of joints, and thought may need to be given as to how to ensure adequate damping. This result should be compared with that of Para. 2 above. Increasing the stiffness of the machine has the further advantage that more accurate work is obtained under all cutting conditions.

7. While this simple theory gives a clear outline of the general story, reality is more complicated: the expression given by Eq. 5.9 is oversimplified, and would need to be modified along the lines of Chap. 4, Para. 5 and 6. Furthermore, linear equations such as are used here give no indication of the amplitude of the chatter. If the damping is treated as being hysteretic, then equations and vector diagrams may readily be modified to give an estimate of amplitude (*ECT* Appendix I.6.1 and *ECT* Appendix III). Furthermore, k_t varies with the depth of cut and the mean speed.

8. The theory is also inadequate for the reason that only one mode of vibration has been considered; where other modes exist with very similar frequencies, coupling can occur, and the tool may have motions along three orthogonal axes and also torsionally about its axes. Nevertheless, the elementary theory gives a clear guide to the machine designer; in particular, it draws attention to the importance of raising the stiffness of the machine.

9. Chatter due to mode coupling. The third type of chatter, which can be described in terms of simple linear vibration theory, is that due to mode coupling; it normally only occurs if the natural angular frequencies in two orthogonal modes, such as ω_x and ω_y above might represent, are almost equal. Taking the case of the lathe, mode coupling could occur if a force in the x direction causes a displacement in the y direction; and a force in the y direction causes a displacement in the x direction when these directions are normal modes of vibration. We can now write

$$x_f = \frac{1}{k_{xy}} \cdot f \tag{5.20}$$

$$y_p = \frac{1}{k_{yx}} \cdot p \tag{5.21}$$

The constants $\frac{1}{k_{xy}}$ and $\frac{1}{k_{yx}}$ are usually known as the *flexibilities* of the system. By Maxwell's reciprocal theorem they must be equal to one another in a linear system.

Since the force f is producing a displacement x_f this is equivalent to adding a force $p_f = k_x \, . \, x_f = \frac{k_x}{k_{xy}} \, . f$ in the x direction. This leads to the equation of motion

$$[M_x\mathrm{D}^2+\lambda_x\mathrm{D}+k_x]x = p+\frac{k_x}{k_{xy}} \, . f \tag{5.22}$$

If, as in Eq. 5.2, $p = \lambda_a \, . \, \mathrm{D}x$
and for simplicity (cf. Eq. 5.9) $f = -k_t y$ (5.23)
Substituting in Eq. 5.22 gives

$$[M_x\mathrm{D}^2 + (\lambda_x - \lambda_a)\mathrm{D}+k_x]x = -\frac{k_x k_t}{k_{xy}} \, . \, y \tag{5.24}$$

Similarly

$$[M_y\mathrm{D}^2+\lambda_y\mathrm{D}+(k_x+k_t)]y = \frac{k_y\lambda_a}{k_{yx}} \, . \, \mathrm{D}x \tag{5.25}$$

Eliminating y from these two equations, and noting that $k_{xy} = k_{yx}$

$$[M_x\mathrm{D}^2+(\lambda_x-\lambda_a)\mathrm{D}+k_x][\mathrm{M}_y\mathrm{D}^2+\lambda_y\mathrm{D}+(k_x+k_t)]x+ \frac{k_x k_y k_t}{k_{xy}^2} \, . \, \lambda_a\mathrm{D}x = 0 \tag{5.26}$$

A similar equation is obtained for y.

These fourth order equations can represent either stable or unstable motion depending on the values of the constants. For the general case of steady sinusoidal motion, the motions in the x and y directions are both sinusoidal but of different amplitudes: their phase relationship is determined by the equations. This leads theoretically to the tip of the tool moving on an elliptical path in the $x-y$ plane. Experimentally this is found to be approximately true.

If the lathe is also assumed to have a third normal mode in the z direction, i.e. along the line of the tool, two further equations may be formed to show the interaction of the x and z, and of the y and z modes: (indeed theoretically all three modes may be simultaneously coupled to give a sixth order equation).

If the z subscript is used to indicate the constants for this mode, and if the chip thickness factor in the z direction is k_s, the equation of motion for coupling between the y and z modes will be

$$[M_y D^2 + \lambda_y D + (k_x + k_t)][M_z D^2 + \lambda_z D + (k_z + k_s)]z + \frac{k_y k_z k_s k_t}{k_{yz}^2} . z = 0 \quad (5.27)$$

which again may represent either stable or unstable motion. A similar equation is obtained for y.

A word of warning is necessary: the normal modes of a lathe are most unlikely to lie along the three axes which have been used for simplicity and convenience.

Mode coupling may sometimes be eliminated from machines by altering the planes of maximum and minimum stiffness in single components such as the tail-stock of a lathe.

10. Magnitude of chatter forces. As may be seen from Fig. 5.3, and Eq. 5.10, the exciting force at resonance approximately equals $2c_y k_y \times$ (Amplitude of Vibration). Suppose that $2c_y = 1/20$, $k_y = 2.10^5$ lbf/in and that the chatter amplitude is 0·001 in. The force needed to maintain steady state chatter is only 10 lbf. This is astonishingly small, and helps to make clear why chatter occurs so easily. It also, however, makes clear why chatter is such an erratic phenomenon. Thus on a hobbing machine chatter occurred when hobbing mild steel. When the material was changed to a steel with approximately twice the Brinell hardness, chatter could no longer be obtained under identical cutting conditions. However, due to the change of material the cutting forces had approximately doubled, and this had led to an increase in the effective damping coefficient for the machine and the cutting process taken together. Since the mean cutting force initially was of the order of 1 000 lbf, this is perhaps not surprising.

11. Chatter in grinding machines. The theory of chatter in grinding machines is similar to that for regenerative chatter in lathes, except that both workpiece and grinding wheel can develop irregular surfaces, and in consequence two separate finite time delays must be considered when setting up the equations. The grinding wheel can also become loaded, and can cut in an irregular manner.

12. Chatter in drilling machines. Chatter in drilling machines results in the drill having a sinusoidal motion superimposed along its axis, and the depth of the cut taken by one flute depends on the cut taken by the one before it. The number of flutes therefore alters the finite time delay involved in regenerative chatter.

13. Chatter in milling and hobbing machines. Chatter is generally regenerative, but analysis has to allow for more than one tooth being in contact simultaneously. In addition to regenerative chatter, serious vibration may be present due to the cutter applying a forcing frequency of once per tooth and also of higher harmonics, the amplitude of the force being dependent on the width and thickness of the metal to be cut.

Chatter-like effects are also observed in hobbing machines due to the cutting forces on the hob being larger than the weight of the hob head opposing them. As a result, the head is lifted off its slides on to its keeps and then dropped back with hob-gash frequency.

14. Chatter in planers and shapers. Both Arnold and regenerative type chatters are commonly met. The conditions met sometimes differ appreciably from those met in turning when only so-called steady state conditions need be considered, since a large part of the cut may be taking place during the period when the chip formation on the cutting edge is still establishing itself.

15. Backlash and stiction. Backlash and stiction in feed mechanisms both have very deleterious effects on the stability of systems. The effect can be estimated by treating them as a form of hysteretic damping, leading to serious phase shifts. Careful attention must be paid in the design of feed mechanisms to avoid both.

16. Chatter due to interaction of machine components. The full range of feedback loops obtainable is immense due to interaction of the cutting process, kinematic train, frame, bearings, tool and table positioning devices. The analysis of this chapter has been confined to three simple cases: the inclusion of components such as those discussed in the next two chapters can greatly increase the complexity of the problem, without affecting the general lines of attack on their solution.

17. Pseudo-chatter in gear shapers. Marks may sometimes be observed on the flanks of gear teeth cut on shapers which have the typical appearance of chatter but which are nevertheless open-loop phenomena. They may be due to three causes:

a. When the tool commences the cut, a step function of force is applied to the frame of the machine, which can oscillate as a result.

b. The cutting force can cause bounce within the backlash if the machine has been allowed to get out of adjustment.

c. Inertia force due to the stroking mechanism, together with the variation in cutting force due to the variation in cutting velocity, appear to cause waves.

18. Problems

1. Some manufacturers leave the core sand in the castings of their machines in order to increase the damping. Consider the effect of this on (*a*) Arnold, (*b*) Regenerative, (*c*) Mode coupled chatter. (See also Chap. 2, Problem 2.)

2. Set up and solve the equations for coupling between two modes if they are such that components of both cutting forces may be resolved into each of the principle modes.

3. Show that for regenerative chatter the permissible width of cut is always proportional to machine dynamic stiffness, regardless of machine speed.

6. Slideways and hydrostatic bearings

1. Coefficient of friction. For lubricated surfaces the curve of coefficient of friction plotted against a dimensionless number comprising Viscosity multiplied by Rubbing Velocity and divided by Specific Load (i.e. load per unit projected area of bearing) is shown in Fig. 6.1. Three sections of it are labelled. In the zone labelled 'Coulomb friction' (the variety met in elementary mechanics) the surfaces are in contact. This gives rise to the so-called 'stiction' forces when surfaces start to move from rest. In the zone marked 'Boundary lubrication' the surfaces are in intermittent contact. In the zone marked 'Fluid Film lubrication' the surfaces are separated by a wedge of oil, and the frictional force varies approximately with the sliding velocity for small changes about any mean value.

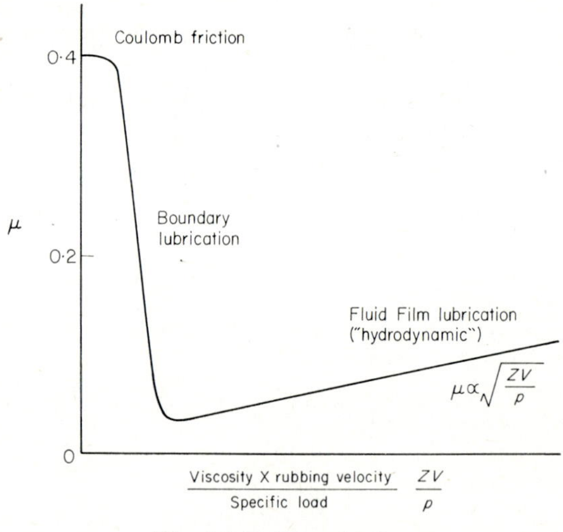

Fig. 6.1. Rubbing friction.

On plain machine-tool slides, friction is normally either Coulomb, after the surfaces have been at rest for a few seconds, or boundary. Both of these conditions make accurate control of a feed system difficult to achieve since a considerable force may be necessary before movement commences; once it does commence, the force necessary may drop away rapidly, thus giving rise to the well known 'stick-slip' phenomena. Troubles due to this may be avoided by using rolling bearings, or hydrostatic bearings. In some circumstances PTFE slides, which have a roughly constant coefficient of friction of 0·04, even when unlubricated, may be used.

2. Rolling bearings. Many forms have been developed recently for machine-tool slides. They tend to have a roughly constant coefficient of friction of about 0·004. Their disadvantages are:
a. They can suffer from 'brinelling' due to vibration of the machine, or if the weight is not taken off the bearing during transport.
b. Their transverse stiffness is often relatively low.
c. They do not have the ability to average out manufacturing errors in the slideways.

3. Hydrostatic bearings. Hydrostatic bearings used in machine tools usually use oil (or air) as their working fluid, but any fluid may in principle be used. An external pump is needed to maintain the fluid film.

Their great advantages are:
a. No stick-slip;
b. When using oil, the coefficient of friction is very small, varying with the velocity: usually about 10^{-5} per in/s;
c. When desired, the pressure can be taken off to give surface to surface contact between the faces, and hence a high coefficient of friction;
d. Wear is negligible: steel ways can be used directly giving economies in design and manufacture of fabricated structures;
e. They have high stiffness in the transverse direction (but low resistance to vibration in the feed direction);
f. They can be made self-compensating to give constant or variable film thickness as may be desired;
g. Errors in the accuracy of the slides are averaged, allowing the use of cheaper surfaces.

Their disadvantages are:

a. The extra complication of an external pump etc. with its additional cost;

b. The possibility of distortion of the machine tool if the oil fed to the bearing is not cool;

c. The low resistance to vibration in the feed direction.

4. Costs. The additional power pack is likely to cost about £100 (at 1969 prices), and additional fitting costs will be incurred. However, the ability to use unhardened surfaces 'as ground', and to avoid the need for special bearing surfaces on steel structures, may well offset the other extra costs, and cause hydrostatic bearings to be economically desirable even when they are not technically essential. For some large rotary bearings, a hydrostatic bearing can be very much cheaper than a crossed roller bearing.

5. Simple theory under static conditions. Fig. 6.2 shows a schematic arrangement of a simple hydrostatic bearing, and Fig. 6.3 the schematic equivalent circuit. The restrictor, of resistance R, may be

a. a simple small bore pipe to give high resistance; or

b. an orifice; or

c. some form of valve which varies the flow resistance in accordance with the bearing pressure.

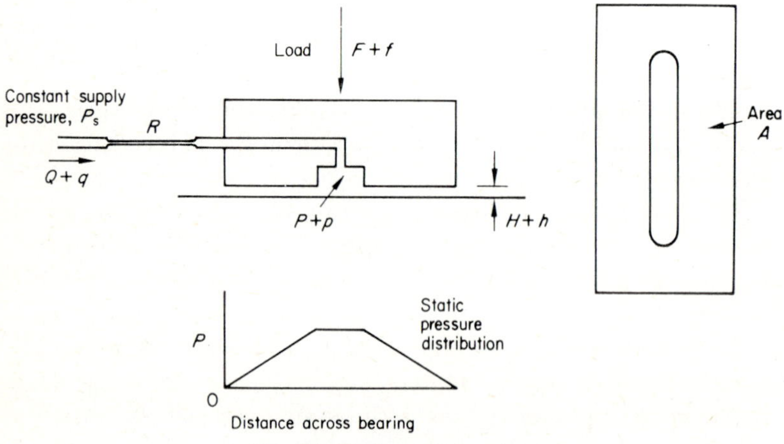

Fig. 6.2. Hydrostatic bearing.

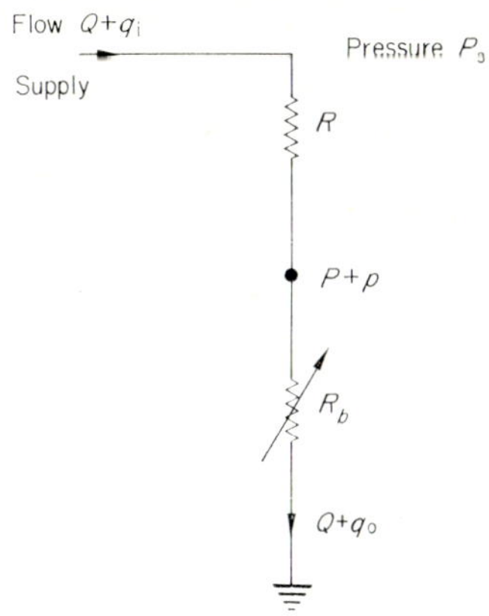

Fig. 6.3. Bearing equivalent circuit.

In the analysis which follows, relationships will be deduced
a. between the load applied to the bearing and the thickness of the oil film;
b. for the stiffness of the bearing both statically and dynamically; and
c. for its damping capacity under alternating loads.

6. Static analysis of bearing with simple restrictor. In Fig. 6.2,
Equating Forces Vertically

$$F = k_1 P \tag{6.1}$$

where k_1 is a constant dependent on the effective area of the pad.
Quantity of Oil Flow Out

$$Q_o = \frac{PH^3}{k_2} \tag{6.2}$$

where k_2 is dependant upon the viscosity of the oil and the geometry of the pad, and is placed in the denominator for later convenience.

In Fig. 6.3,

$$Q_o = \frac{P}{R_b} \tag{6.3}$$

$$\therefore R_b = \frac{k_2}{H^3} \tag{6.4}$$

Quantity of Oil Flow In

$$Q_i = \frac{P_s - P}{R} \tag{6.5}$$

But for continuity of flow, $Q_o = Q_i$, so that

$$P = \frac{P_s R_1}{R+R_1} \tag{6.6}$$

Substituting in Eq. 6.1,

$$F = k_1 \frac{P_s R_b}{R+R_b} = \frac{k_1 k_2 P_s}{RH^3+R_2} \tag{6.7}$$

Note that theoretically as $H \to 0$, $F \to k_1 P_s$: due to lack of flatness of the surfaces, this may not be attained in practice. Now the stiffness of the bearing may be defined as $\frac{dF}{dH}$ and

$$\frac{dF}{dH} = -\frac{3k_1 k_2 R P_s H^2}{(RH^3+k_2)^2} \tag{6.8}$$

This has a maximum where $\frac{d^2F}{dH^2} = 0$,

i.e. where

$$2H(RH^3+k_2)^{-2} - H^2 \,.\, 2 \,.\, 3RH^2(RH^3+k_2)^{-3} = 0$$

or

$$3R^2H^6+(3k_2-1)RH^3-k_2 = 0 \tag{6.9}$$

Substituting from Eq. 6.4 gives $R_b = 2R$. For safety, designs usually make R_b lie between $\frac{1}{2}R$ and R.

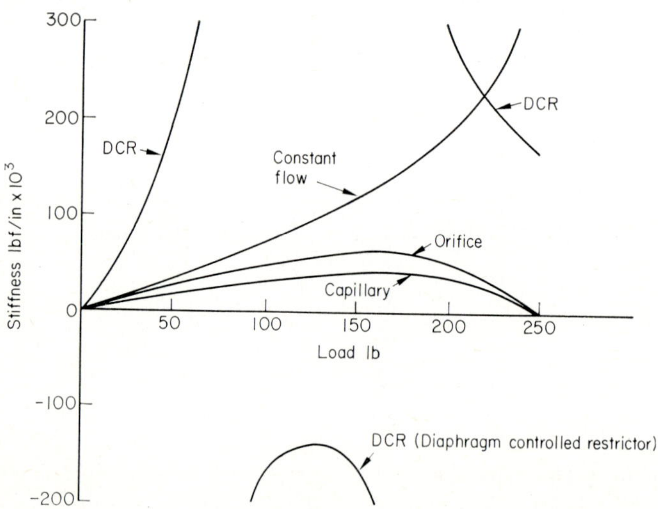

Fig. 6.4. Typical bearing stiffness curves.

The sort of stiffness curve to which this gives rise in a typical bearing is shown in Fig. 6.4.

7. Static analysis of bearing with orifice. An orifice has a non-linear characteristic, since the flow across it is proportional to the square root of the pressure drop. Linearizing the system to make simple analysis possible gives equations identical with those above. However, orifices give stiffer bearings (see Fig. 6.4) but are little used since they must be very small and tend to get blocked. They also cause the bearing to be dependent on the viscosity, and hence the temperature of the oil.

8. Static analysis with variable geometry flow valve. A valve such as that shown in Fig. 6.5 with a diaphragm controlled restrictor (DCR) can be designed to give a pressure drop which is a function of the bearing pocket pressure.

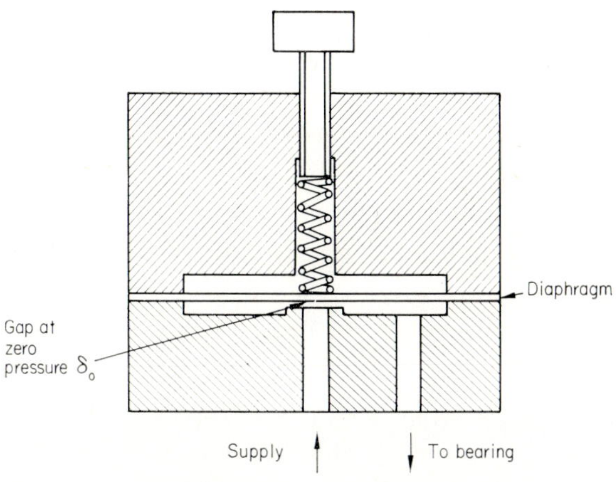

Fig. 6.5. Diaphragm controlled restrictor.

The resistance of the valve R may be written

$$R \propto \frac{constant}{\left(\delta_0 + \frac{P}{k}\right)^3}$$

$$= \frac{R_0}{\left(1 + \frac{P}{k_4}\right)^3} \tag{6.10}$$

where R_0 is the resistance at zero pressure and

$$k_4 = k\delta_0$$

Substituting in Eq. 6.5

$$\frac{PH^3}{k_2} = \frac{P_s - P}{R}$$

$$= \frac{(P_s - P)\left(1 + \frac{P}{k_4}\right)^3}{R_0}$$

Since $F = k_1 P$,

$$H^3 = \frac{k_1 k_2}{R_0} \frac{\left(P_s - \frac{F}{k_1}\right)\left(1 + \frac{F}{k_1 k_4}\right)^3}{F} \tag{6.11}$$

Since H must be $\geqslant 0$, $P_s \geqslant \frac{F}{k_1}$

Differentiating to obtain the stiffness $\frac{dF}{dH}$

$$\frac{R_0}{k_1 k_2} \cdot 3H^2 \frac{dH}{dF} = -\frac{P_s}{F^2}\left(1 + \frac{F}{k_1 k_4}\right)^3 + \frac{3}{k_1 k_4} \times \frac{\left(P_s - \frac{F}{k_1}\right)\left(1 + \frac{F}{k_1 k_4}\right)^2}{F}$$

$$\therefore \frac{dF}{dH} = \frac{3R_0}{k_1 k_2} \cdot \frac{F^2 H^2}{\left(1 + \frac{F}{k_1 k_4}\right)^2} \cdot \frac{1}{\left[\left(-P_s - \frac{P_s F}{k_1 k_4}\right) + \left(\frac{3P_s F}{k_1 k_4} - \frac{3F^2}{k_1^2 k_4}\right)\right]} \tag{6.12}$$

Inspecting the bottom RH square bracket suggests that it might be $+^{ve}$, $-^{ve}$ or *zero*. This last possibility is of particular interest, since if this term is zero, the RHS becomes infinite, and the bearing has infinite stiffness. What are the conditions for this to be possible?

Rearranging the bracket gives

$$-\left[+P_s - \frac{2P_s F}{k_1 k_4} + \frac{3F^2}{k_1^2 k_4}\right]$$

For this to $= 0$,

$$F = \frac{+k_1 P_s \pm k_1 \sqrt{P_s^2 - 3k_4 P_s}}{3} \tag{6.13}$$

As the load on the bearing changes, the stiffness will in general be infinite for two values. Since $k_4 = k\delta_0$, these can be adjusted by altering either the spring-rate k of the valve, or the initial setting δ_0.

The stiffness curve obtainable with such bearings is shown in Fig. 6.4

9. Static analysis of constant flow valve. An interesting special case of the above analysis occurs if, by using a variable rate spring,

$R = R_0 - k_3 P$, where k_3 is a constant. (6.14)

Substituting in Eq. 6.3

$$Q_0 = \frac{P}{R_b} = \frac{P_s - P}{R}$$

$$= \frac{P_s - P}{k_3\left(\frac{R_0}{k_3} - P\right)}$$

This is a constant if $\frac{R_0}{k_3} = P_s$

Designing the valve to give this condition makes

$$P = \frac{R_b}{k_3}$$

and

$$F = \frac{k_1}{k_3} R_b = \frac{k_1 k_2}{k_3} \cdot \frac{1}{H^3} \qquad (6.15)$$

$$\text{Stiffness of the bearing} = \frac{dF}{dH} = -\frac{3k_1 k_2}{k_3} \cdot \frac{1}{H^4} \qquad (6.16)$$

which increases with bearing load. This is shown in Fig. 6.4.

10. Static oil flow characteristics. The oil flow Q through the bearing is plotted against P in Fig. 6.6 for the commonest arrangements:

a. With a capillary restrictor the bearing clearance reduces as the load increases, giving a linear drop in flow;

b. An orifice as restrictor gives a somewhat flatter curve;

c. A 'constant flow' valve manages to maintain this condition over a wide range;

d. A valve designed to give infinite stiffness implies a constant gap;

increasing P should therefore give a linear increase in flow, as shown by the dotted line; practical solutions tend to be as shown, due to limitations on supply pressure eventually allowing the bearing to close.

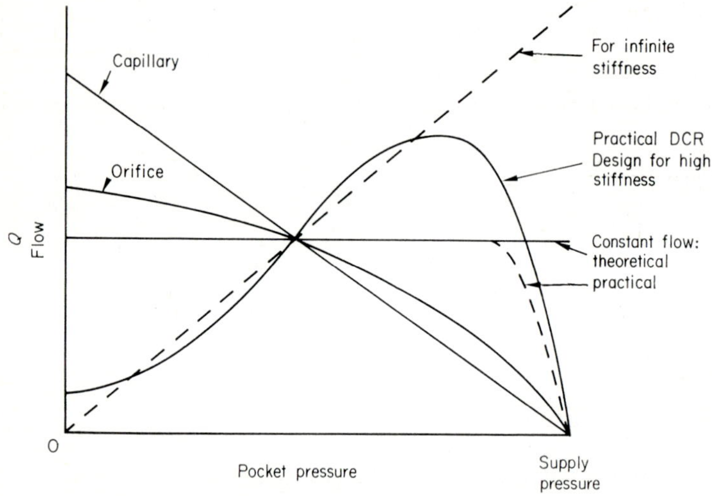

Fig. 6.6. Bearing flow requirements.

11. Dynamic analysis for simple restrictor and compressible fluid. Suppose that for some reason the faces of the bearing are moving so as to alter the gap. Typically this would be due to a changing load, or variations of pressure within the bearing. The rate at which oil will be stored within the gap is $A \,.\, \mathrm{D}h$ where h is a small change in H. At any instant the rate of flow into the bearing will not be equal to the rate of flow out of it, although the average values must be the same.

In addition, if the oil is not thoroughly de-aerated, or if the volume of oil is high, compressibility of the fluid can play an important part. Exact analysis is difficult, but for oil bearings, so long as cavitation is not occurring, the assumptions which follow are reasonable.

Due to the change of pressure within the bearing, resulting from change in external loading, the fluid will be compressed and change its volume. If the original mean volume is V (approximately equal to the volume of the oil pockets plus half the volume of the gap in the mean position) and K is the Bulk Modulus, then the rate of storing due to compressibility is $KV \,.\, \mathrm{D}p$, where p is a small change from the mean pressure P. (The Bulk Modulus for oil is shown in Fig. 7.11; for gases, $k = \gamma P$, where γ is the ratio of the specific heat and P is the absolute pressure.) From Fig. 6.2 and 6.3

Flow in—Flow out = Volume of oil stored,

$$\therefore q_i - q_o = A \,.\, \mathrm{D}h + KV \,.\, \mathrm{D}p \tag{6.17}$$

As before $Q_o = \dfrac{PH^3}{k_2}$

Differentiating gives $\mathrm{d}Q_o = \dfrac{H^3}{k_2} \,.\, \mathrm{d}P + \dfrac{3PH^2}{k_2} \,.\, \mathrm{d}H$

or writing small changes as small letters,

$$q_o = \frac{H^3}{k_2} \cdot p + \frac{3PH^2}{k_2} \,.\, h$$

Now $q_i = -\dfrac{p}{R}$

Substituting in Eq. 6.17 gives

$$-\frac{p}{R} - \left(\frac{H^3}{k_2} \cdot p + \frac{3PH^2}{k_2} \,.\, h\right) = A \,.\, \mathrm{D}h + kV \,.\, \mathrm{D}p$$

But $F = k_1 P$ as before

$$\therefore \left[\frac{k_1 k_2 A}{3H^2 F}\mathrm{D}+1\right]h = -\frac{k_1(k_2+RH^3)}{3H^2FR} \cdot \left[\frac{3KVH^2FR}{k_1(k_2+RH^3)}\mathrm{D}+1\right]p \tag{6.18}$$

Assuming that inertia forces may be neglected, since the motion is very small, $f = k_1 p$ and

$$\left[\frac{k_1 k_2 A}{3H^2 F}\mathrm{D}+1\right]h = -\frac{(k_2+RH^3)}{3H^2FR}\left[\frac{3KVH^2FR}{k_1(k_2+RH^3)}\mathrm{D}+1\right]f \tag{6.19}$$

This may conveniently be rewritten

$$[T\mathrm{D}+1]h = -\mu[T_\rho \mathrm{D}+1]f \tag{6.20}$$

where T is a time constant, T_ρ is a derivative time constant (*ECT* Section 3.3) and μ is the sensitivity (or gain) of the unit.

12. Dynamic analysis considering inertia of system. If the mass supported on the bearing is M, the exact relationship between f and p is given by

$$M\mathrm{D}^2 h = -f + k_1 \cdot p \tag{6.21}$$

Substituting this in Eq. 6.18 gives

$$[T\mathrm{D}+1]h = -\mu[T_\rho\mathrm{D}+1].[M\mathrm{D}^2h+f]$$

or

$$[\mu MT_\rho\mathrm{D}^3+\mu M\mathrm{D}^2+T\mathrm{D}+1]h = -\mu[T_\rho\mathrm{D}+1]f \tag{6.22}$$

Whether this represents a stable bearing or not may be determined from Routh's criterion. For stability,

$$T\,.\,\mu M \geqslant \mu MT_\rho$$

or

$$T \geqslant T_\rho \tag{6.23}$$

The term T_ρ contains the Bulk Modulus and Volume of fluid in the bearing, whereas T does not.

13. Dynamic stiffness and damping capacity of hydrostatic bearings. For a system such as that discussed in Para. 11 when the force is sinusoidal of angular frequency ω, Eq. 6.20 may be rewritten

$$[T\mathrm{D}+1]h = -\mu[T_\rho\mathrm{D}+1]f\sin\omega t \tag{6.24}$$

The solution of this equation (*ECT*, Chapters 1 and 3) can be seen in the vector diagram Fig. 6.7 (cf. *ECT*, Section 3.3) which gives the dynamic stiffness,

$$\frac{f}{h} = \frac{\sqrt{\omega^2T^2+1}}{\mu\sqrt{\omega^2T_\rho^2+1}} \tag{6.25}$$

which tends, at high frequencies, to $\dfrac{T}{\mu T_\rho}$, a constant.

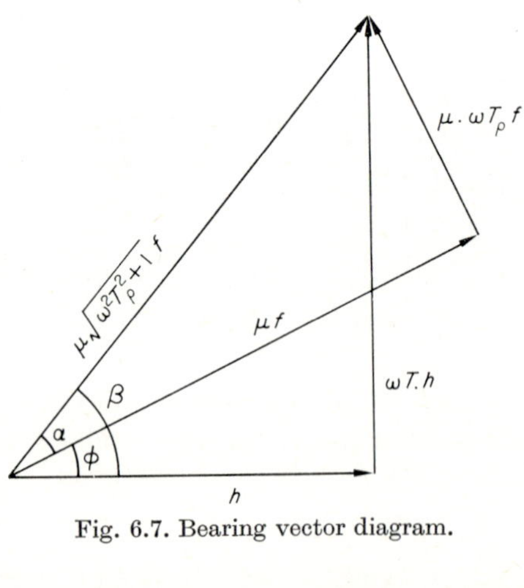

Fig. 6.7. Bearing vector diagram.

The energy dissipated by such a system may be found by noting that it is given by the time integral of the component of force resolved in the direction of the velocity. From the figure

$$\phi = \beta - \alpha = \tan^{-1}\omega T_\rho - \tan^{-1}\omega T \tag{6.26}$$

$$\begin{aligned}
\text{Work per cycle} &= \int_0^{\frac{2\pi}{\omega}} (f\sin\omega t)[h\omega\cos(\omega t - \phi)]\mathrm{d}t \\
&= \pi f h \sin\phi \\
&= \pi f h(+\cos\alpha\sin\beta - \sin\alpha\cos\beta) \\
&= \pi f h\left(+\frac{1}{\sqrt{\omega^2T_\rho^2+1}}\cdot\frac{\omega T h}{\mu f\sqrt{\omega^2T_\rho^2+1}} - \frac{\omega T_\rho}{\sqrt{\omega^2T^2+1}}\cdot\frac{h}{\mu f\sqrt{\omega^2T^2+1}}\right) \\
&= \frac{\pi\omega h^2}{\mu}\left(\frac{+T-T_\rho}{\omega^2T_\rho^2+1}\right)
\end{aligned}$$

But $h^2 = \mu^2 f^2\left(\dfrac{\omega^2T_\rho^2+1}{\omega^2T^2+1}\right)$

$$\text{Work/cycle} = \frac{\mu\pi\omega f^2(+T-T_\rho)}{\omega^2T^2+1} \tag{6.27}$$

At high frequencies the damping tends to be very small, or negligible, since it is varying almost inversely as the frequency. This is because bearing deflection is negligible.

With variable flow valves when the stiffness is infinite, negligible amounts of energy can be dissipated.

14. Hydrostatic nuts. The same principles as used in slideway bearings can be applied to the design of nuts to go on lead screws. Difficulties, however, may be encountered due to the fact that effective bearing areas are very small, so pressures and flow rates are extremely high. Alignment problems with hydrostatic nuts are particularly troublesome, so that it is usually more satisfactory to use a recirculating ball nut and screw.

15. Air and gas bearings. The analysis starting in Para. 11 gives an approximate solution for air bearings as well as for oil lubricated ones, but the exact analysis is extremely complicated due to the air being expanded by the work done on it in the bearing.

When designing air bearings, reference should be made to one of the standard works. The coefficient of friction which is achieved with air bearings is effectively zero and loads per unit area of the bearing of the order of $\frac{1}{4}$ to $\frac{1}{3}$ of the supply pressure can be carried. Stability problems are frequently encountered.

16. Problem

1. Consider the effect on the analysis of this chapter of delays in oil lines such as those discussed in Chap. 7, Para. 15.

7. Hydraulic servo controls

1. Hydraulic position controls are extensively used in machine tools both for positioning tables and slides, and also for moving the tool in copy-lathes. Well designed, they are rugged, accurate, and free from both backlash and deadzone. They can also have very high response rates, and time constants of the order of 0·001 s can readily be achieved with them, even when forces of the order of a ton have to be produced. Fig. 7.1 shows a typical application for a copy-lathe.

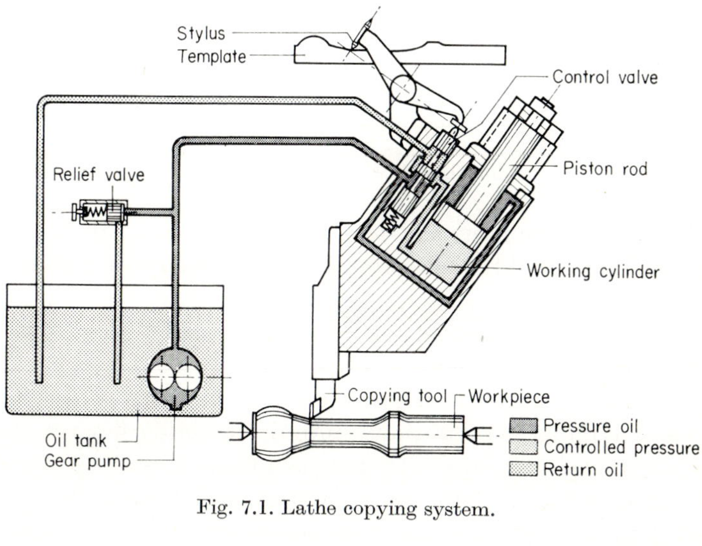

Fig. 7.1. Lathe copying system.

Despite their virtues, hydraulic components can on occasion behave in ways which at first sight appear to be irrational, and are often difficult to diagnose. A thorough understanding of their fundamental modes of operation, and their theoretical analysis, can often lead to the diagnosis of the problems and hence to their solution—or indeed to their avoidance at the design stage. Only rarely are figures quoted in this chapter, since any minor changes in design or in manufacture, or even in the pressure at which a system is run, can lead to major changes in port coefficients and design 'constants'.

For movement of the tool in copy-lathes, a piston-type mechanism is usually the best solution. For table positioning a choice exists between using a ram; using a hydraulic motor with nut and lead screw, either of the recirculating ball-nut type, or of the hydrostatic nut type; or using motor and rack drive. The range over which each is appropriate is shown approximately in Fig. 7.2. The reason for the choice of lbf in as a scale will appear in Para. 2.

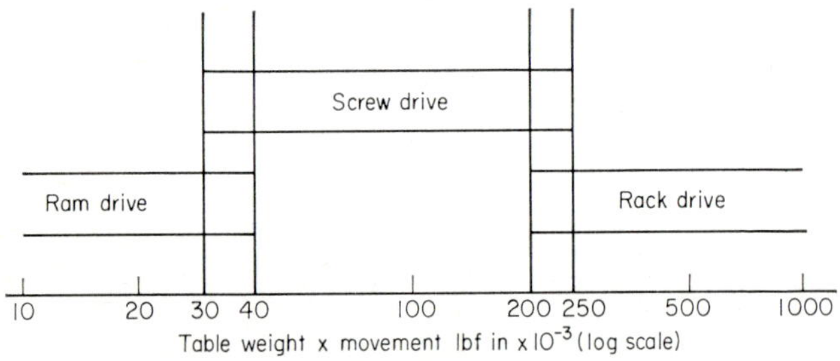

Fig. 7.2. Drive mechanism ranges.

Both hydraulic rams and also hydraulic motors can hunt, and the reasons are often obscure. The more common reasons are discussed later.

Normal electric motors are not used because they typically have time constants of the order of $\frac{1}{4}$ s.

Hydraulic position controls often form part of open or closed loops containing electrical and optical measuring devices. These are not discussed in this book, but their characteristics must often be taken into account. Since closed control loops frequently reduce to the equivalent of three simple time delays, a note has been added in Appendix II on the stability of such systems.

2. Hydraulic positioning cylinders. Oil is compressible, the bulk modulus usually being in the range of 1×10^5 psi to a maximum of 2×10^5 psi, the value being very dependent on the degree of aeration as is shown in Fig. 7.11. A positioning cylinder filled with oil therefore acts as a spring, and if used to position a table, trouble may occur unless table clamps are fitted to prevent motion when the table is meant to be held in a fixed position.

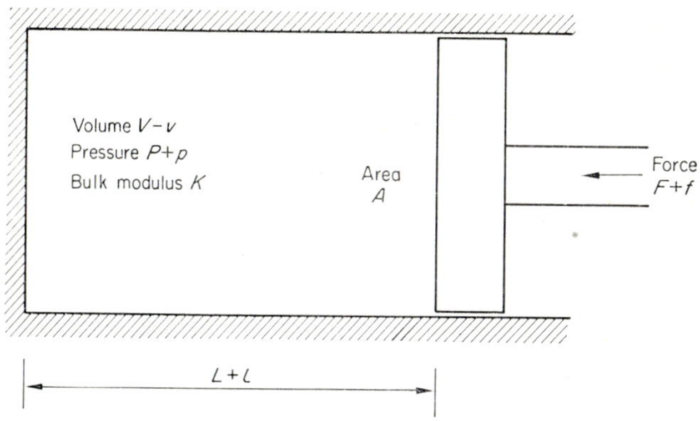

Fig. 7.3. Hydraulic ram parameters.

The spring constant of a single acting cylinder may be derived using the nomenclature of Fig. 7.3. For compressible fluid in a cylinder, $p = -K\left(\frac{v}{V}\right)$, since increasing the pressure reduces the volume. Therefore change in force $f = pA = -KA\,\frac{Al}{AL} = -\frac{KA}{L}\,.\,l$

$$\therefore \text{Spring Constant of cylinder} = \frac{KA}{L} \tag{7.1}$$

If the mass attached to the piston is M,

$$\text{Natural Frequency of undamped system} = \sqrt{\frac{KA}{L}\cdot\frac{1}{M}} \tag{7.2}$$

This is the reason why, for a given cylinder, LM is a useful criterion. This is shown by Fig. 7.4.

3. Elementary analysis of hydraulic servomechanisms. Much time can be saved in the analysis of hydraulic servomechanisms if they are initially tackled using the simplest possible assumptions; only if these fail to solve the problem need recourse be made to more refined ones. In general, equations for (*a*) geometry, (*b*) flow and (*c*) force balance will be sought. Block diagrams of the system can also be useful, but often the same mechanical element will occur in two different blocks on the diagram, since it is fulfilling

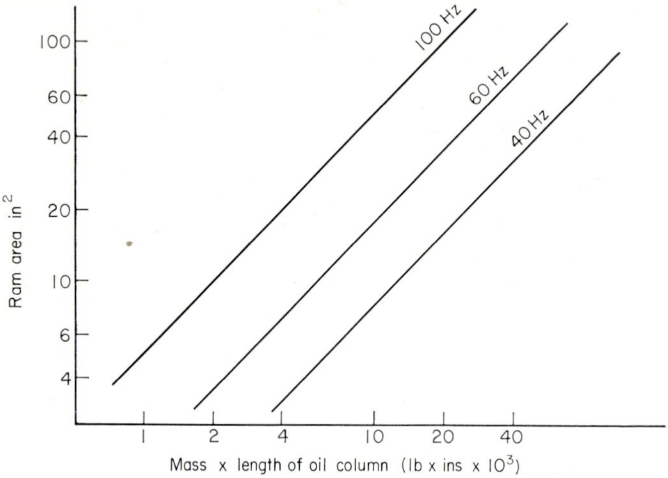

Fig. 7.4. Variation of ram area with mass, natural frequency, and stroke for single acting cylinder.

Ram area (in^2) $= 5{\cdot}12 \times 10^{-7} \times$ Frequency^2(Hz^2) $\times$ Mass(lb) $\times$ Length of oil column (in).

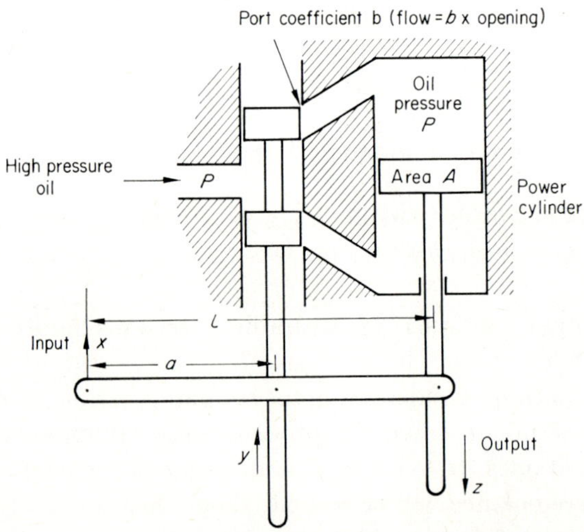

Fig. 7.5. Simple hydraulic servo. x, y, z represent small displacements from the mean position.

two different functions. The systems which are analysed below have all caused trouble to the authors in practical applications.

4. Simple position control. In Fig. 7.5 the assumptions are made that:

a. the lands exactly cover the ports in the mean position;

b. no backlash exists at the joints;

c. the oil is incompressible;

d. the rate of flow through the valves is proportional to the amount that they are open, and independent of any pressure drop across them;

e. no leakage;

f. the areas of the two sides of the piston are equal.

While none of these is exactly true, the results obtained apply in very many cases sufficiently accurately.

The *flow equation* is found by equating the rate of change of volume of the cylinder to the rate of flow through the valve

$$A\,.\,\mathrm{D}z = by \tag{7.3}$$

From *geometry*,

$$y = \frac{l-a}{l}\cdot x - \frac{a}{l}z \tag{7.4}$$

Eliminating y:

$$\left[\frac{A}{b}\cdot\frac{l}{a}\mathrm{D}+1\right]z = \left[\frac{l-a}{l}\right]x \tag{7.5}$$

or

$$[T\mathrm{D}+1]z = \mu_{\mathrm{s}}x \tag{7.6}$$

when T is a time constant, and μ_{s} is the Sensitivity or Gain. With careful design, T may be of the order of 1 to 2 milliseconds. The block diagram for this unit is shown in Fig. 7.6.

5. Passive proportional+derivative device. For the device shown in Fig. 7.7,

From geometry, $z = y + \dfrac{l}{a}(x-y)$ (7.7)

From force balance, $\lambda \mathrm{D}y = k(x-y)$ (7.8)

Eliminating y, $\left[\frac{\lambda}{k}\mathrm{D}+1\right]z = \left[r\,.\,\frac{\lambda}{k}\mathrm{D}+1\right]x$ where $r = \frac{l}{a}$

Putting $r\frac{\lambda}{k} = \rho$, the Derivative Time constant, gives

$$\frac{z}{x} = \frac{[\rho\mathrm{D}+1]}{\left[\frac{\rho}{r}\mathrm{D}+1\right]} \tag{7.9}$$

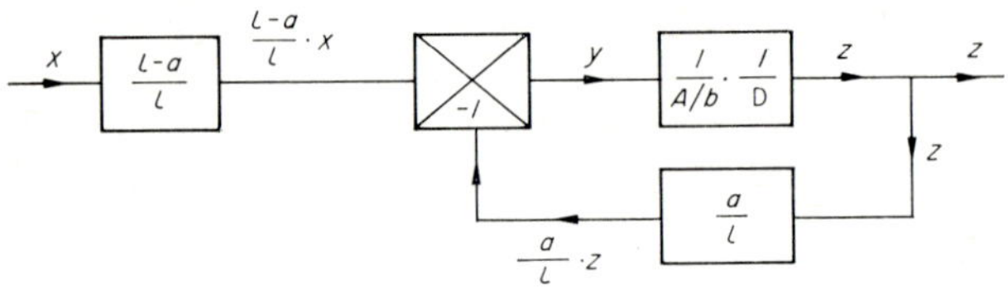

Fig. 7.6. Block diagram representing simple hydraulic servo.

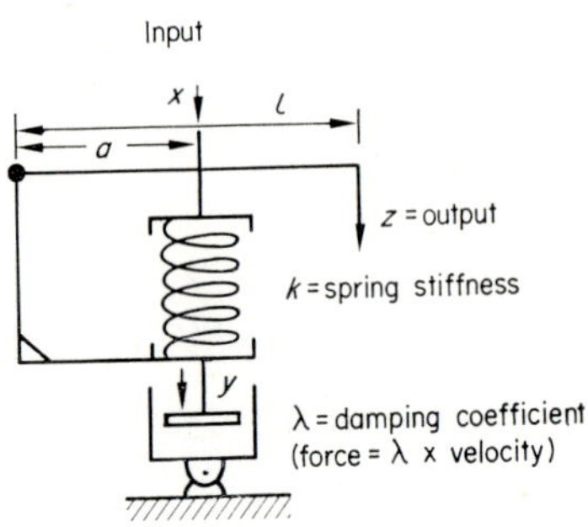

Fig. 7.7. Passive device giving proportional and derivative action.

Care must be taken when designing units of this sort to make sure that the oil does not cavitate, and the suction should not be more than about 6 lbf/in^2. Approximate compensation for the change of the oil viscosity with temperature can be obtained by making the piston of a material which has a greater coefficient of expansion than the cylinder.

6. Proportional+derivative servomechanism. For the device shown in Fig. 7.8

Flow Equation $A\mathrm{D}z = b(x-y)$ (7.10)

Force Balance $\lambda\mathrm{D}y = k\left(\frac{z}{r}-y\right)$ (7.11)

Eliminating y gives $\left[\frac{\lambda}{k}\cdot\frac{rA}{b}D^2+\frac{rA}{b}D+1\right]z = r\left[\frac{\lambda}{k}D+1\right]x$ (7.12)

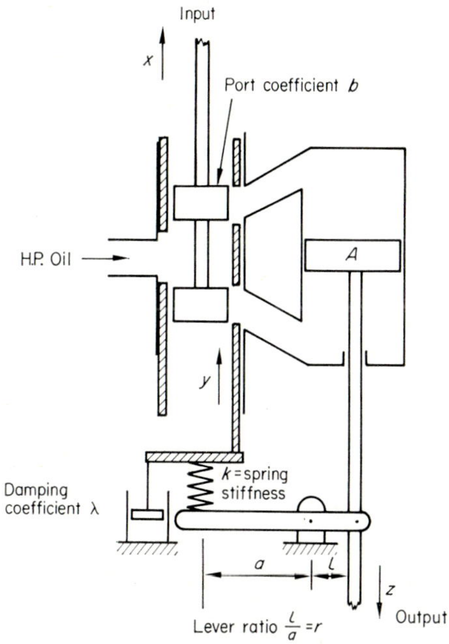

Fig. 7.8. Hydraulic servo giving proportional and derivative action.

7. Proportional+integral servomechanism. For the device shown in Fig. 7.9,

Flow Equation, $ADz = b(x-y)$ (as in 7.10)

Force Balance, $ky = \frac{\lambda}{r}\left(Dz-Dy\right)$ (7.13)

Eliminating y, $\left[\frac{rA}{rA+b\frac{\lambda}{k}}\cdot\frac{\lambda}{k}D+1\right]z = \frac{rb}{rA+b\frac{\lambda}{k}}\cdot\frac{\lambda}{k}\left[1+\frac{1}{\frac{\lambda}{k}D}\right]x$

(7.14)

or

$$[TD+1]z = \mu\left[1+\frac{1}{T_\gamma D}\right]x \qquad (7.15)$$

where $T_\gamma = \frac{\lambda}{k}$ is called the Integral Time Constant.

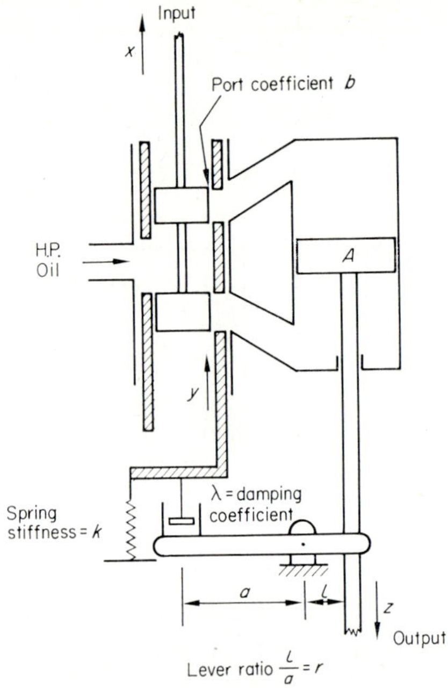

Fig. 7.9. Hydraulic servo giving proportional and integral action.

8. Analysis of instability in hydraulic servomechanisms. Instability and oscillation of hydraulic servomechanisms may be caused by at least seven different mechanisms, alone or in combination. Usually one reason is found to predominate in any given mode of oscillation. Some of these oscillations can be predicted from linear theory, some from non-linear theory, and some qualitatively.

9. Oscillations due to oil compressibility and power piston inertia. Fig. 7.10 shows the same simple system as that discussed in Para. 4. Assume in this case that the oil is compressible, that the output piston has inertia and that the rate of flow of oil through the valve ports is no longer proportional only to their opening, but also to the pressure difference across them. The consequence of the second assumption is that the difference in oil pressure across the power piston must be taken into account.

The compressibility of the oil depends very largely on the degree of aeration of the oil, as is shown in Fig. 7.11. The assumption of compressibility means that the mass of oil in the cylinder depends not only on the volume of the cylinder, but also on the pressure in it. The assumption is further made that the pressure in the cylinder never drops so far that cavitation occurs in the oil. This may not necessarily be true.

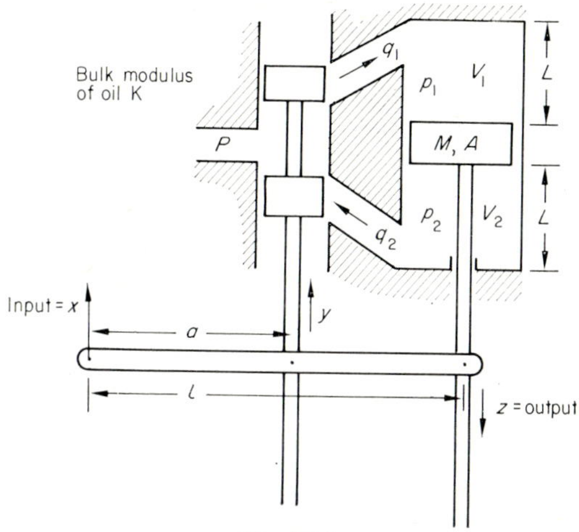

Fig. 7.10.

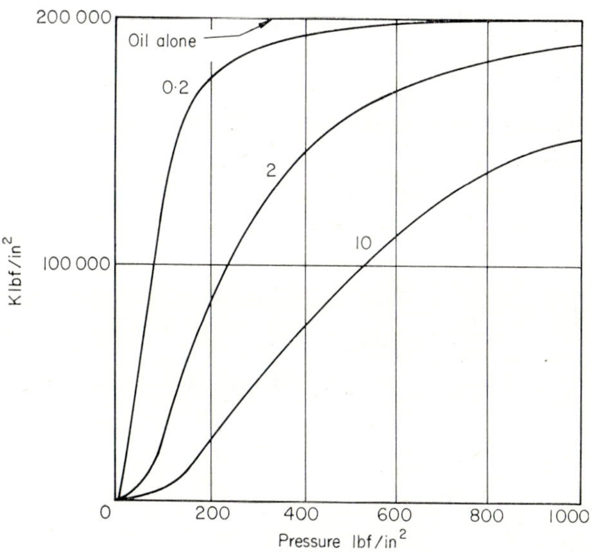

Fig. 7.11. Bulk modulus of oil. Curves show variation of bulk modulus with pressure for several percentage ratios of free air volume to liquid volume at atmospheric pressure. (From *I. Mech. E. Conference on Servomechanisms*, 1953, p. 49.)

The equation of motion for the power piston may be written

$$MD^2z = (p_1-p_2)A \tag{7.16}$$

The flow equations into the top and bottom of the cylinder have two terms, one due to the change in volume of the cylinder, one due to the flow needed to compensate for the compressibility of the oil. Due to compressibility $p = -K\left(\frac{v}{V}\right)$ where v is the change in volume due to the pressure, K the bulk modulus.

Differentiating gives $Dp = -\left(\frac{K}{V}\right)Dv$. (The $-^{ve}$ sign is because increase of pressure reduces the volume.) The flow equation into the top half is

$$q_1 = \frac{V_1}{K}Dp_1 + ADz \tag{7.17}$$

and into the bottom

$$q_2 = -\frac{V_2}{K}Dp_2 + ADz \tag{7.18}$$

(noting that $ADz = -DV$).
The flow through the ports is also determined by a second set of equations dependent on their geometry; it is assumed to be directly proportional both to the amount which they are open, and also to the pressure drop across them. (These assumptions are only approximately true.)
Assuming that the valve flows are symmetrical this gives

$$q_1 = k_1y - k_2p_1 \tag{7.19}$$

and

$$q_2 = k_1y + k_2p_2 \tag{7.20}$$

where k_1 and k_2 are constants.
For small movements about mid-position, assume that $V_1 = V_2 = V$.
From Eq. 7.17 and 7.19

$$\left[\frac{V}{K}D + k_2\right]p_1 = -ADz + k_1y \tag{7.21}$$

and from Eq. 7.18 and 7.20

$$\left[\frac{V}{K}D + k_2\right]p_2 = +ADz - k_1y \tag{7.22}$$

Subtracting Eq. 7.22 from Eq. 7.21

$$\left[\frac{V}{K}\mathrm{D}+k_2\right](p_1-p_2) = -2A\mathrm{D}z+2k_1y \tag{7.23}$$

From geometry,

$$y = \frac{l-a}{l}x-\frac{l}{a}z \tag{7.24}$$

Eliminating (p_1-p_2) and y from Eq. 7.16, 7.23 and 7.24 gives

$$M\mathrm{D}^2z = \frac{A}{\frac{V}{K}\mathrm{D}+k_2}\left[-2A\mathrm{D}z+\frac{2k_1}{l}\{(l-a)x-az\}\right]$$

$$\therefore\left[\frac{MV}{AK}\cdot\frac{l}{a\,.\,2k_1}\mathrm{D}^3+\frac{M}{2A}\frac{l}{a}\frac{k_2}{k_1}\mathrm{D}^2+\frac{lA}{ak_1}\mathrm{D}+1\right]z = \frac{l-a}{a}x \tag{7.25}$$

For stability, from Routh's rule,

$$\frac{k_2}{k_1} > \frac{a}{l}\frac{V}{A}\cdot\frac{1}{K} \text{ or noting that } V = A\times L$$

$$\frac{k_2}{k_1} > \frac{a}{l}\cdot\frac{L}{K} \tag{7.26}$$

The ratio k_2/k_1 is the ratio of the port coefficients for the flow rates with change of pressure and change of opening. Neither of these values is in fact a constant, and this may well lead to complications compared with this simple analysis.

Similar equations are also found for the system:

a. if the piston works against a spring;
b. if the piston has viscous damping;
c. if oil leaks past the piston due to the pressure difference across it.

10. Reaction forces on valves. Experiments with a two-dimensional valve system shown in Fig. 7.12 (i) show that two main flow patterns are possible. Where the valve is only cracked open, the main flow through the valve is in the direction shown at *1*, and is parallel to the wall. Eddies also exist in the 'dead' spaces. This implies that the fluid changes its direction of flow through 90° with a corresponding change of momentum. The resultant increase of force on the valve is shown in zone *1* of Fig. 7.12 (iii). The discharge is shown in Fig. 7.12 (ii), and in the region *1* the flow coefficient of the valve (i.e. the ratio of the actual flow to the maximum possible theoretical flow) is about 0·9.

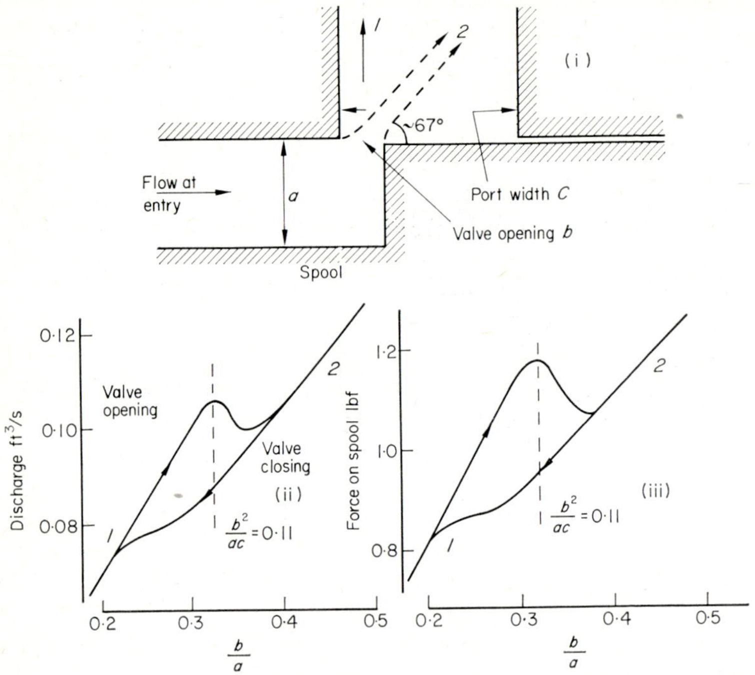

Fig. 7.12. Spool valve forces. *1*, is direction of flow for small openings; *2*, is direction of flow for large openings.

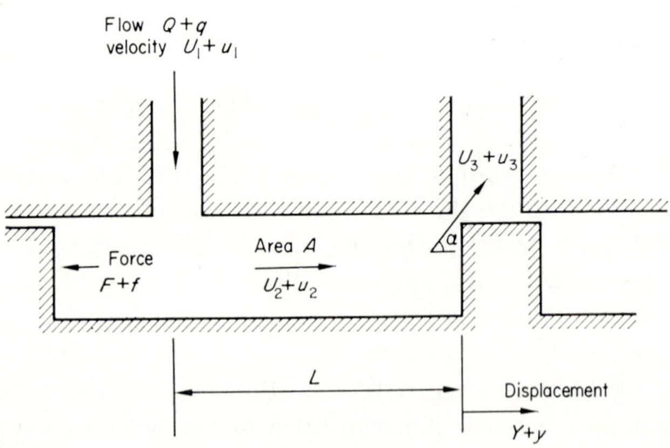

Fig. 7.13. Valve reaction forces.

When, however, the valve is opened further, the flow pattern changes until the main flow is at an angle of about 67° to its original direction, and the flow coefficient is only just over 0·6. The momentum change is correspondingly smaller, as is shown in zone *2* of Fig. 7.12 (iii) by the change in slope of the force curve.

As is shown in Fig. 7.12 (ii) and (iii) the change from one flow regime to the other occurs at different points when opening and shutting the valve. This has important results as will be shown later.

11. Reaction forces on valves when in motion. The forces on the valve may be analysed approximately for either of the steady flow patterns discussed above.

In Fig. 7.13 if the net resultant force tending to close the valve is $F+f$, from the momentum equation for *steady state*

$$F\mathrm{d}t = \rho Q\,\mathrm{d}t\,(U_2-0)+\rho Q\,\mathrm{d}t\,(U_3\cos\alpha-U_2)+\rho LA\,\mathrm{d}U_2 \qquad (7.27)$$

Momentum change at entry+Momentum change at exit+
Momentum change due to change in rate of flow.

But $Q = AU_2$
and $Q = b_1U_3Y$
and $Q = b_2Y\sqrt{P}$
} where b_1 and b_2 are constants: and making the usual assumptions for flow through an orifice.

Substituting and tidying up,

$$F = (\cos\alpha)b_2Y\sqrt{P}\,.\,\frac{b_2}{b_1}\sqrt{P}+\rho Lb_2\frac{\mathrm{d}}{\mathrm{d}t}(Y\sqrt{P})$$

Differentiating, neglecting second order terms, and writing f for dF etc.

$$f = \frac{b_2^2}{b_1}\rho(\cos\alpha)[P\,.\,y+Y\,.\,p]+b_2\rho L\left[\sqrt{P}\,.\,\mathrm{D}y+\frac{Y}{2\sqrt{P}}\mathrm{D}p\right] \qquad (7.28)$$

The change in closing force is thus seen to depend not only on the change, but also on the rate of change, both of opening and of pressure. This may be written

$$f = [k_3\mathrm{D}+k_1]y+[k_4\mathrm{D}+k_2]\,p \qquad (7.29)$$

12. Reduction of reaction forces on valves. Two main methods are used, both of which depend on causing the oil to have equal momentum changes on entering and leaving the valve. One solution is to shape the face of the spool so that the oil tends to cling to the boundary of the port as it does for small openings. This is shown in Fig. 7.14.

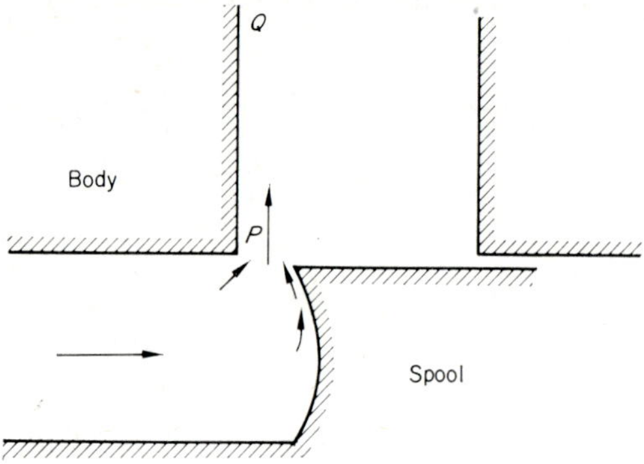

Fig. 7.14. Valve port flows. Shaped spool causes oil to flow parallel to PQ.

Another solution is to have thick valve stems which stop the oil following a diagonal path through the valve, and force it to change direction through 90° both at inlet and exit.

13. Oscillations due to valve reaction forces. For practical purposes the analysis given in Para. 11 is often adequate, and when this is so, oscillations due to valve closing forces may be treated by simple linear theory. When, however, both flow patterns exist, non-linear analysis may be necessary, and one example of this is given in Para. 16.

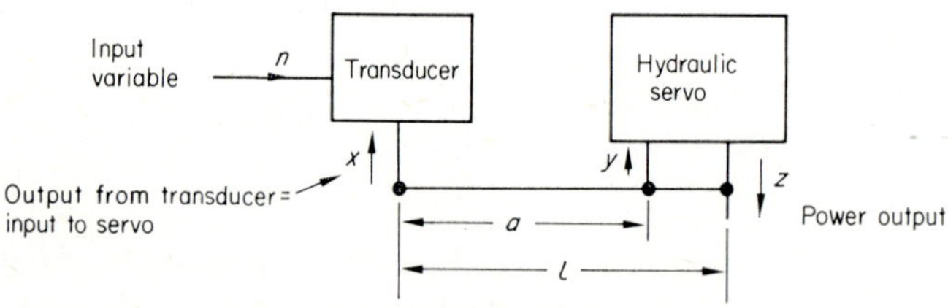

Fig. 7.15.

14. Oscillations due to valve reaction forces and non-rigid inputs. In the system of Fig. 7.15 the hydraulic servo is assumed to behave as a simple time delay, as analysed in Para. 4. The transducer is assumed to be a damped spring-mass system, and the input to the servo is assumed to be subject to a closing force dependent on its position and velocity as in Eq. 7.29.

For the input valve,

$$[T_1 D+1]y = \mu_1 f \tag{7.30}$$

For the servo,

$$[T_2 D+1]z = \frac{l-a}{l} x \tag{7.31}$$

From geometry

$$y = \frac{l-a}{l} . x - \frac{a}{l} . z \tag{7.32}$$

For the transducer,

$$\left[\frac{D^2}{\omega_0^2}+\frac{2c}{\omega_0}D+1\right]x = \mu_g . n - \frac{(l-a)}{l} f \tag{7.33}$$

Where $n = 0$, eliminating f, x and y, and putting $\frac{l-a}{l} = \mu_2$,

$$\left[\frac{D^2}{\omega_0^2}+\frac{2c}{\omega_0}D+1\right][T_2 D+1]z + \frac{\mu_2^3}{\mu_1}\left[\frac{T_2}{\mu_2}D+1\right][T_1 D+1]\, z = 0 \tag{7.34}$$

The stability may be investigated in the usual ways: it will be very sensitive to changes in μ_2.

15. Oscillations due to valve reaction forces and supply line resonance. In a system such as that of Fig. 7.16 when the valve spool is assumed to be attached to an input which is flexible, or else contains backlash, oscillations can occur at high frequency due to the natural frequency of the input valve corresponding with the resonance frequency of oil in the supply line.

a. Equation of motion of the spool. From Eq. 7.29

$$[mD^2+\lambda D]y = -[k_3 D+k_1]y - [k_4 D+k_2]p \tag{7.35}$$

$$\left.\begin{array}{l}\text{Inertia}\\ +\\ \text{Damping forces}\end{array}\right\} = \left\{\begin{array}{l}\text{Closing forces due to change of displacement}\\ +\\ \text{Closing forces due to change of pressure}\end{array}\right.$$

b. Behaviour of oil in supply line. Consider first a section of uniform pipe as in Fig. 7.17, where no restrictions are placed on the nature of the small pressure and flow changes. A small portion of oil of length dx is a distance x along the pipe; a short time dt later the initial boundaries of the oil have moved to $(x+y)$ and $[(x+\mathrm{d}x)+(y+\mathrm{d}y)]$

$$\therefore \text{Increase in length} = \mathrm{d}y = \frac{\mathrm{d}y}{\mathrm{d}x}\mathrm{d}x$$

$$\therefore \text{Volumetric strain} = \frac{A\dfrac{\mathrm{d}y}{\mathrm{d}x}\mathrm{d}x}{A\,\mathrm{d}x} = \frac{\mathrm{d}y}{\mathrm{d}x}$$

If K is the bulk modulus,

$$p = -K\frac{\mathrm{d}y}{\mathrm{d}x} \tag{7.36}$$

since increasing the pressure decreases the volume. Rigid walls to the pipes are assumed.

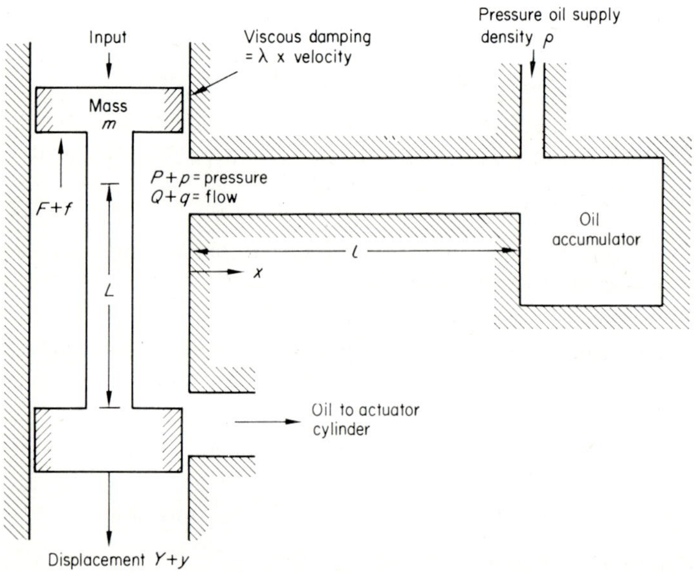

Fig. 7.16.

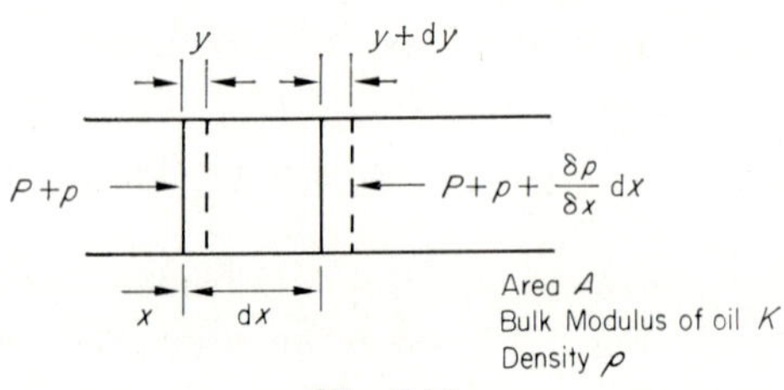

Fig. 7.17.

The equation of motion of the element, neglecting friction, is

$$A\rho\,\mathrm{d}x\,\frac{\partial^2 y}{\partial t^2} = Ap - A\left(p + \frac{\partial p}{\partial x}\,\mathrm{d}x\right)$$

$$\therefore\ \rho\,\frac{\partial^2 y}{\partial t^2} = -\frac{\partial p}{\partial x} \tag{7.37}$$

Substituting from Eq. 7.36

$$\therefore\ \rho\,\frac{\partial^2 y}{\partial t^2} = +K\,\frac{\partial^2 y}{\partial x^2} \tag{7.38}$$

This is the classical wave equation representing a wave being propagated with a velocity $\sqrt{K/\rho}$.

(This analysis may be compared with that of Chap. 9, Para. 1.)

The problem may now be simplified by considering only the case when trouble occurs, i.e. when the system is in oscillation. Fourier analysis allows steady oscillation to be treated as the sum of a number of sinusoidal terms.

If this equation is solved for steady sinusoidal motion, the solution is of the form

$$y = B\sin(\Omega x - \psi)\sin(\omega t - \varphi)$$

If x = constant = X, then $\sin(\Omega x - \psi)$ is a constant, and the equation represents the velocity variations with time at X. If t = const., then $\sin(\omega t - \phi)$ is a constant, and the equation represents the velocity variations at any given moment as x changes.

Differentiating,

$$\frac{\partial^2 y}{\partial t^2} = -B\omega^2\sin(\Omega x - \psi)\sin(\omega t - \varphi) \tag{7.39}$$

and

$$\frac{\partial^2 y}{\partial x^2} = -B\Omega^2\sin(\Omega x - \psi)\sin(\omega t - \varphi) \tag{7.40}$$

$\therefore$ Dividing Eq. 7.39 by Eq. 7.40 and substituting from Eq. 7.38 gives

$$\omega^2 = \frac{K}{\rho}\,\Omega^2 \tag{7.41}$$

$$\therefore\ y = B\sin\left(\omega\sqrt{\frac{\rho}{K}}\,x - \psi\right)\sin(\omega t - \varphi) \tag{7.42}$$

This repeats itself down the pipe each time $\left(\omega\sqrt{\frac{\rho}{K}}\right)x$ increases by 2π, and hence the wavelength of pressure and sound waves in the pipe is $\left(\frac{2\pi}{\omega}\right)\sqrt{\frac{K}{\rho}}$.

$$\text{Now } p = -K\frac{\partial y}{\partial x} = -B\omega\sqrt{K\rho}\cos\left(\omega\sqrt{\frac{\rho}{K}}x-\psi\right)\sin(\omega t-\phi) \tag{7.43}$$

and the change from the mean rate of oil flow down the pipe is given by

$$q = A\frac{\partial y}{\partial t} = A\,.\,B\omega\sin\left(\omega\sqrt{\frac{\rho}{K}}x-\psi\right)\cos(\omega t-\phi)$$

$$\therefore \frac{p}{q} = -\frac{\sqrt{K\rho}}{A}\cot\left(\omega\sqrt{\frac{\rho}{K}}x-\psi\right)\tan(\omega t-\phi) \tag{7.44}$$

For an open-ended pipe or a great enlargement at $x = l$, as in Fig. 7.16 a small change in p must produce a very large change in q.

$$\therefore 0 = -\frac{\sqrt{K\rho}}{A}\cot\left(\omega\sqrt{\frac{\rho}{K}}l-\psi\right)\tan(\omega t-\phi)$$

This must be true for all values of t,

$$\therefore -\cot\left(\omega\sqrt{\frac{\rho}{K}}l-\psi\right) = 0$$

$$\therefore \omega\sqrt{\frac{\rho}{K}}l-\psi = \pm\frac{\pi}{2} \text{ for the fundamental frequency}$$

$\therefore$ At the valve, where $x = 0$

$$\frac{p}{q} = \frac{\sqrt{K\rho}}{A}\tan\omega l\sqrt{\frac{\rho}{K}}\,.\,\tan(\omega t-\phi) \tag{7.45}$$

When oscillation is occurring at the valve, this is satisfied for $p = p_0\sin\omega t$ if

$$q = p_0\cos\omega t\,.\,\frac{A}{\sqrt{K\rho}}\cot\omega l\sqrt{\frac{\rho}{K}} \tag{7.46}$$

[*Note.* If at $x = l$ the end is closed, for any value of p, $q = 0$,

$$\therefore \frac{p}{q} \to \infty$$

and

$$\therefore \psi = \omega l\sqrt{\frac{\rho}{K}} \text{ etc.]}$$

c. *Condition for valve to be in steady oscillation.* Now

$$q = k_5 y + k_6 p \tag{7.47}$$

Substituting in this Eq. from Eq. 7.46 for steady sinusoidal motion, when

$$q = \mathrm{j}\frac{A}{\sqrt{K\rho}}\cot \omega l\sqrt{\frac{\rho}{K}} \cdot p$$

gives

$$\mathrm{j}\frac{A}{\sqrt{K\rho}}\cot \omega l\frac{\rho}{K} \cdot p = k_5 y + k_6 p$$

Substituting in Eq. 7.35 for p gives

$$\frac{[m\mathrm{D}^2 + (\lambda + k_3)\mathrm{D} + k_1]}{[k_4\mathrm{D} + k_2]}y = \frac{k_5}{\mathrm{j}\dfrac{A}{\sqrt{K\rho}}\cot \omega l\sqrt{\dfrac{\rho}{A}} - k_6}y \tag{7.48}$$

This awkward expression may be investigated by plotting two vector loci for $y = y_0 \sin \omega t$, one for the LHS and one for the RHS of the equation, and considering where they intersect. If the intersection occurs at the same frequency on each curve, then the assumed stable harmonic oscillation is possible. Inspection will show the effect of changing the sensitivity (gain) of one side or the other, and of changing any of the other constants. This is shown in Fig. 7.18.

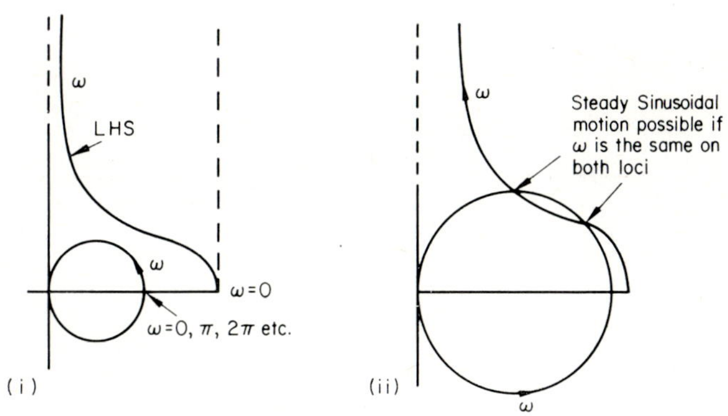

Fig. 7.18. (i) Stable system, (ii) Potentially oscillatory system.

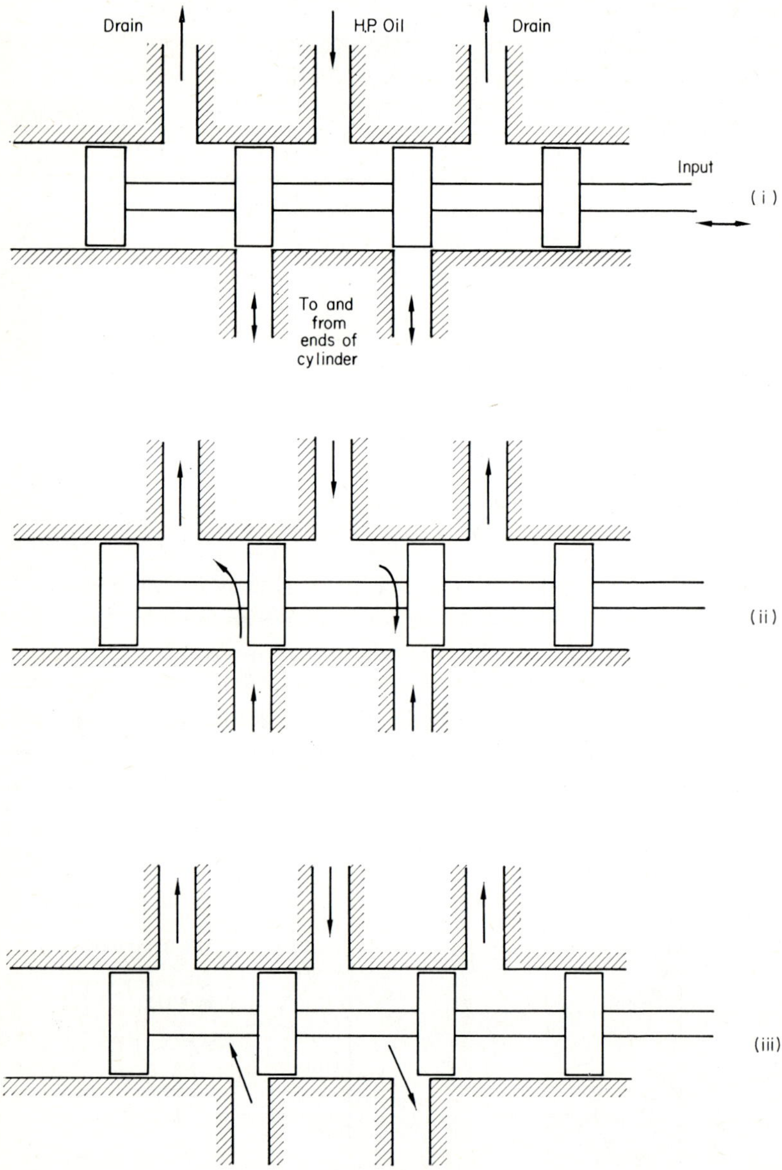

Fig. 7.19. Valve flows.

16. Phase plane analysis of oscillations in four-land spool valve due to hysteretic valve reaction forces. The ordinary four-land spool valve is shown in Fig. 7.19 (i) in the closed position. When it starts to open, as in Fig. 7.19 (ii), the flow through it tends to have a pattern which is approximately as shown on the figure; this, however, shows only a two dimensional approximation to a very complicated three dimensional process. The flow of oil both into and out of the spool tends to be along the faces of the lands; since the axial momentum change as the oil enters and leaves the spool is identical, no axial forces occur on the valve due to the oil flow.

When the valve is wide open, the flow is as shown in Fig. 7.19 (iii). At the left hand end, axial momentum must be destroyed; in the middle it must be created; in both cases this leads to an axial force tending to close the valve.

The change-over from one flow pattern to the other usually occurs at different positions when the valve is opening and when it is shutting, with the result that a graph of closing force plotted against valve opening is obtained which may be idealized and approximated by a curve such as is shown in Fig. 7.20 (i). The hysteresis loop found has interesting implications for the valve behaviour as may be shown neatly in the Phase Plane in Fig. 7.20 (ii). The elementary theory of the Phase Plane is given in Appendix I.

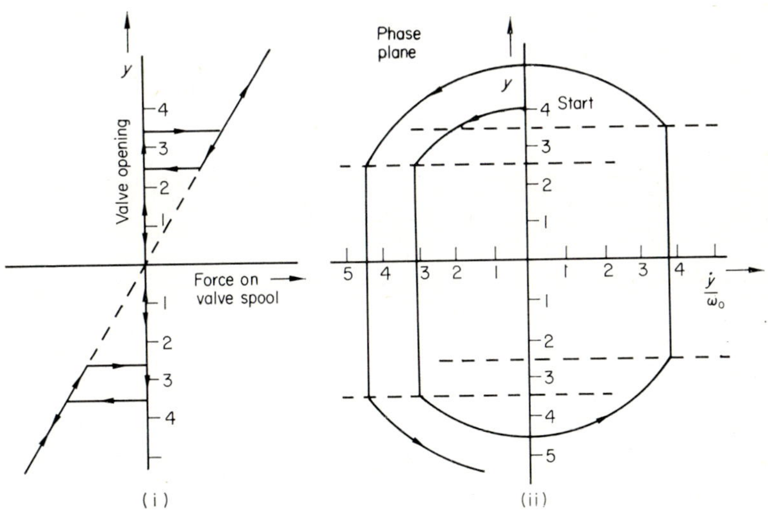

Fig. 7.20. Valve instability.

In Fig. 7.20 the spool is assumed to have mass but no damping, and to be free to move subject only to valve forces. This might occur in a system with backlash. If initially the spool is held with a displacement of four units and then released, the spool will try to centre due to oil forces proportional to the displacement acting like a spring. When the displacement reduces to $2\frac{1}{2}$ units the oil force drops to zero, and the spool continues to move at constant velocity due to its mass (no damping!) until the spool has moved to $3\frac{1}{2}$ units in the other direction. Oil forces now take charge once more, and the cycle continues in this manner with constantly increasing amplitude. In fact damping must occur, and this would result in a limiting value of the amplitude occurring. However, during experiments with valves of this sort, the amplitude has built up so far that a free spool has shot out of the valve altogether.

Even when backlash is not present, oscillations can occur if the input is not rigid, since this merely adds a constant spring term. The analysis proceeds in a similar manner, but the natural frequency changes from ω_{01} to ω_{02}, and thus the scale along the abscissa from $\dfrac{\dot{y}}{\omega_{01}}$ to $\dfrac{\dot{y}}{\omega_{02}}$ and back again. The result of this on the form of the diagram is shown in Fig. 7.21.

For the sake of clarity in the diagram, the stiffnesses have been chosen to give $\omega_{01} = 2\omega_{02}$. Starting as before with a displacement of $y = 4$ and $\dot{y} = 0$, the point in the phase plane starts at *0* and

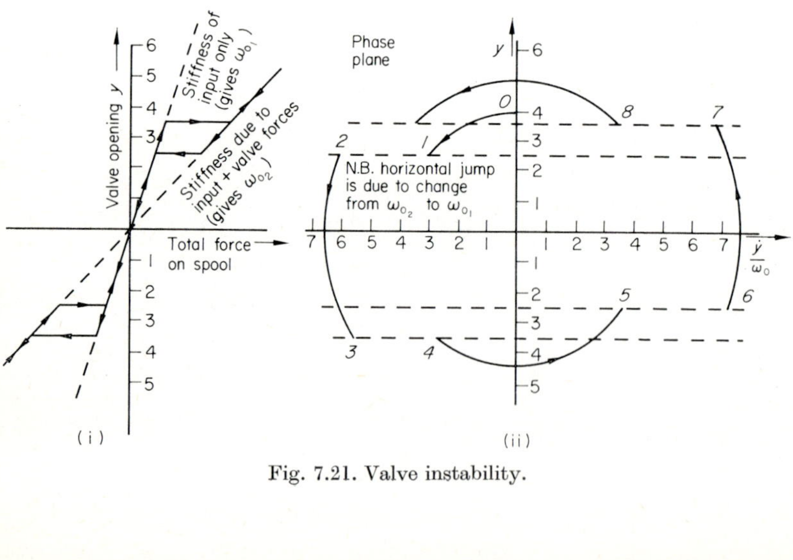

Fig. 7.21. Valve instability.

moves on the arc of a circle with centre at the origin to the point *1*. Here the effective stiffness, and hence the frequency, reduces, so the scale on the abscissa of the phase plane changes, and the point representing the valve in the phase plane jumps from *1* to *2*. The point then moves on another arc of a circle with centre at the origin until it reaches *3*. Due to increase of the effective frequency, the point jumps from *3* to *4*; and so on. These jumps do *not* represent changes in velocity.

17. Valve stiction and hydraulic lock. Non-linear vibrations may also occur in the system due to the valve sticking at one end of its travel until large enough forces build up for it to move suddenly to the other extreme position, when it sticks again in a similar manner.

The most common reason for this, at low pressures, is hydraulic lock. One way of reducing, but not of eliminating, the trouble is the use of oil grooves round the piston. Keeping the components revolving relative to one another cures the trouble completely.

Mathematical analysis of the pressure distribution round a nominally cylindrical piston in a cylindrical sleeve shows that three cases must be considered.

a. The piston and the cylinder are both perfectly cylindrical; the pressure distribution will, given time, force the piston hard against one side of the cylinder, and if they are well finished they will 'wring' together.

b. The cylinder is perfectly cylindrical, but the piston is slightly tapered (1 part in 10 000 is enough, 1 in 1 000 is ample) with the higher pressure at the wider end of the piston; the effect is the same as for (*a*).

c. The cylinder is perfectly cylindrical, but as before the piston is tapered; if the higher pressure is at the smaller end of the piston, the piston is self-centring and will be free in the cylinder. The three cases are shown exaggerated in Fig. 7.22.

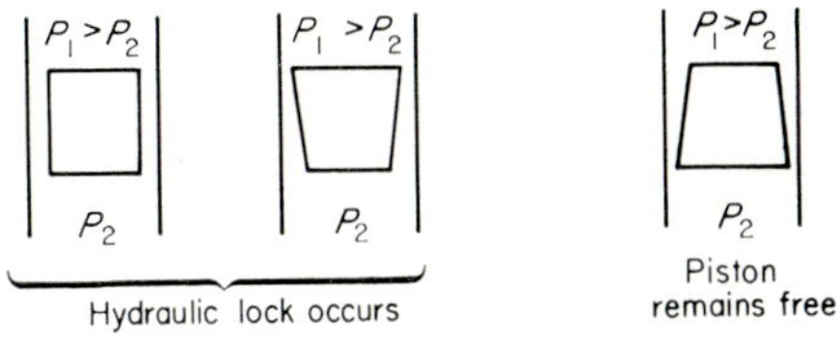

Fig. 7.22. Schematic representation of 'cylindrical' piston in cylinder.

This analysis is sometimes found to break down at pressures over about 1 500 lbf/in² because the pressure distribution may itself so distort the shapes of the components that effectively the taper gets reversed from what it was before pressure was applied. Similar effects are also often due to dirt in the oil.

18. Oscillations due to dirt. A very high proportion of all troubles in hydraulic circuits are due to dirt in the oil. Not even a new barrel of oil should be assumed to be clean.

Both linear and also rotary servomotors may sometimes be observed to jump impulsively at regular intervals. This may be caused by fine dirt in the oil gradually silting up the very small valve openings which are needed to compensate for the leakage past the cylinder or actuator. When the flow stops the actuator will slowly move until it causes the feedback mechanism to open the valve far enough to clear itself of dirt and to reposition itself. This cycle can proceed with surprising regularity.

19. Compensation for unknown forces by force feedback. The forces acting on the power piston may be very non-linear in nature due to hydraulic lock on the piston or due to the nature of the forces opposing motion. As an example, the forces on the cutting tool of a servo copying lathe are non-linear.

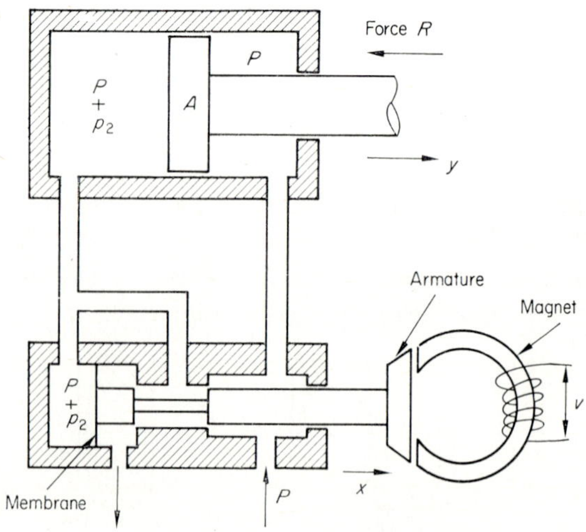

Fig. 7.23. Load compensation.

For a theoretically perfect system, complete compensation for such forces can be obtained with the system shown in Fig. 7.23 if the components are suitably designed. Due to oil compressibility and leakage, perfect compensation is in fact impossible, but very good compensation can usually be achieved.

Equation of Motion for Piston. Suppose that an unwanted force R exists on the output piston, then

$$p_2 A = M\mathrm{D}^2 y - R \tag{7.49}$$

Equation for Valve Displacement in terms of applied forces. When k_2, k_3, etc., are suitable constants,

$$x = k_2 v + k_3 p_2 \tag{7.50}$$

Flow Equation through control valve. Neglecting oil compressibility

$$A\mathrm{D}y = k_4 x - k_5 p_2 \tag{7.51}$$

Substitute from Eq. 7.50 in Eq. 7.51 to eliminate x

$$A\mathrm{D}y = k_2 k_4 v + (k_3 k_4 - k_5) p_2 \tag{7.52}$$

Substitute from Eq. 7.49 in Eq. 7.52 to eliminate p_2

$$A\mathrm{D}y + (k_5 - k_3 k_4)\frac{1}{A}[M\mathrm{D}^2 y - R] = k_2 k_4 v \tag{7.53}$$

If, by suitable design, we make

$$k_5 - k_3 k_4 = 0, \tag{7.54}$$

then

$$y = \frac{k_2 k_4}{A} \cdot \frac{1}{\mathrm{D}} v \tag{7.55}$$

i.e. a perfect integral control is obtained with the effect of unwanted forces eliminated.

This may be modified to give proportional control by adding a feedback link to give

$$x = k_2 v + k_3 p_2 - k_6 y \tag{7.56}$$

and

$$\therefore [A\mathrm{D} + k_4 k_6] y = k_2 k_4 v + (k_3 k_4 - k_5) p_2 \tag{7.57}$$

$$\therefore \left[\frac{A}{k_4 k_6}\mathrm{D} + 1\right] y = \frac{1}{A} \cdot \frac{k_2}{k_6} v \tag{7.58}$$

Note that a mechanical input would need to act on the armature through a spring, so as to apply a force. Alternatively, the whole valve body might be moved.

20. Oscillations due to size of oil passages between valve and power cylinder. Occasionally trouble may occur due to the oil passages between valve and power cylinder containing a large volume of oil, which is compressible. If the passages are long—occasionally long pipes may separate valve and actuator—the inertia of the oil may be appreciable, and a finite time delay of considerable magnitude may occur in the propagation of the pressure wave. The analysis of these troubles is similar to those already given, but is usually not very rewarding. A quick solution may sometimes be obtained by replacing rigid metal pipes with reinforced rubber pressure pipes which introduce a large measure of damping.

A pipe with too small a bore will lead to high velocities which involve high kinetic energies, and hence high apparent inertias. The design problems introduced by inertia and compressibility are therefore contradictory.

21. Oscillations due to backlash and other manufacturing and design errors. Backlash causes the output to fail to follow the input whenever the direction of motion reverses; it is generally destabilizing, since it leads to indeterminacy in a system. Overlap of valve ports leads to a similar indeterminacy, but in practice a small amount is sometimes found to be stabilizing in a system, partly because it reduces the sensitivity to very small perturbations about the mean position, partly because an accurately lapped valve has a very slight effective opening due to the difference in valve and bore diameters. Care should be taken in the design to preload any part of the system in which wear can occur so that backlash is taken up automatically.

22. Spring-mass systems with stick-slip but negligible inertia. Flow valves are sometimes so arranged that they are positioned by oil acting on a piston opposed by a spring. A good estimate of the effect of stick-slip in a system of this sort when it is being moved sinusoidally can be obtained by means of a very simple graphical construction, which is shown in Fig. 7.24. The sinusoidal input is shown in the bottom LH corner, and above it is shown the force characteristic of the system plotted against displacement. In the top RH corner the output is plotted by the graphical

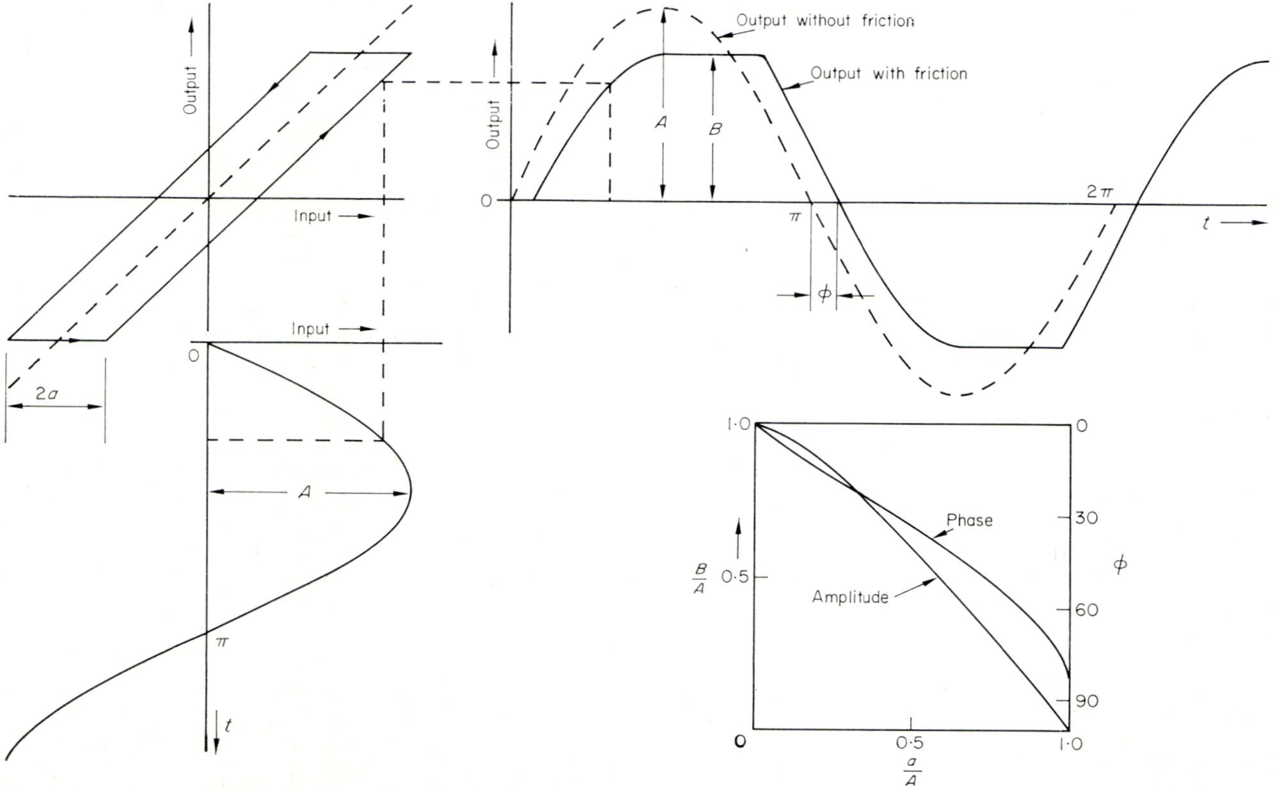

Fig. 7.24. Coulomb Friction effects.

construction for which one set of construction lines is shown. If this construction is repeated for different amplitudes of input, the curves shown in the bottom RH corner are obtained.

Inspection of the shape of the output shows that a considerable second harmonic is also present, as well as higher harmonics, but these can usually all be neglected as a first approximation.

An analytical technique which gives the same information is known as 'describing functions'.

23. Problem

1. Draw block diagrams for all the systems discussed.

8. Vibration dampers

1. When investigating the problem of reducing vibration in a machine, the decision must be made as to whether a problem is one of vibration isolation or vibration absorption. The main troubles encountered in machine tools are usually:

a. External vibrations from other machines, transmitted through the foundations. On small rigid machines, vibration isolation of the machine from the floor is usually the best solution. On large machines, the machine relies on its foundation for alignment, so the whole foundation block must be isolated.

b. Vibrations from auxiliaries on the machines. If auxiliaries such as hydraulic pumps or coolant pumps give trouble they can usually be vibration isolated from the machine.

c. Vibrations from the cutting process. Interrupted cuts such as occur in milling or hobbing give forced vibrations which cannot be isolated; their effect must be absorbed or reduced by increasing machine stiffness.

d. Chatter arises from lack of machine dynamic stiffness and may require the use of absorbers. Isolators are of no help.

e. Roughness in machine main drive. Unsuitable bearings, misalignments or eccentricities in the main drive of a machine may give inadmissible vibration, particularly with grinding machines. The cause of vibration must be eliminated.

f. Motor vibrations. If the motor is axially coupled to its load, it can usually be vibration isolated from the structure of the machine. When a motor is belt-driving a machine, the force in the belt is *not* small in relation to the weight of the motor, so that if the motor is mounted on conventional isolators, full load belt tension would move the motor a large distance. It is not possible to give a general solution to this problem.

The techniques of vibration isolation are relatively straightforward and are dealt with in standard textbooks. When isolation cannot be used, the possibilities of absorbers should be considered.

2. As machine-tool structures are improved the static stiffness of a machine structure is usually increased but the dynamic stiffness is little altered; high values of Q, the dynamic amplification factor at resonance, tend to be found.

Often the most economical way of improving a machine with high resonance peaks is to increase the damping rather than the static stiffness. Attempts have been made to improve dynamic stiffness by deliberately designing relative movement, and hence friction losses, into the main structure using joints which are not welded or shafts made of shrunk sleeves. These methods are fundamentally bad since they invariably reduce the static stiffness of the machine and the resultant machine is both statically and dynamically inferior to a properly designed machine with energy absorption.

3. Three fundamentally different principles may be used to reduce vibration levels:

a. Tuned auxiliary masses vibrating to give a force opposing the motion.

b. Energy absorption extracting energy from the system and dissipating it as heat.

c. Types (*a*) and (*b*) are combined in tuned damped absorbers which give a compromise between the characteristics of the two types.

d. Active systems using a servo-controlled system to oppose motion.

The types of vibration absorber normally used are:

a. Tuned undamped absorber. This can only be used where there is a fixed forcing frequency (with a.c. motors, 25, 50 and 100 Hz may all occur) or where the forcing frequency is associated with the speed of rotation of a shaft. Such absorbers are of no use for chatter.

b i. Impulse dampers. These, as the name suggests, rely on impacts to absorb energy from the system. The simplest design consists of partly filling a casting or hollow shaft with sand or ball bearings. Though this approach is extremely cheap and reasonably effective for rough work, it cannot be used for accurate work since a vibration amplitude of the order of 0·001 in (25μ) is needed before impulse dampers will start to work. Since analysis of these dampers is very difficult, trial and error is the best method of setting them up in practice.

ii. Friction coupled inertias with Coulomb damping. The most common use for these is on torsionally vibrating systems where a friction-type or Lanchester damper can be fitted. They are not usually used in machine tools because their performance is limited even when on optimum settings; their performance is amplitude dependant; and they require good, intelligent maintenance, since the friction linings wear.

iii. Viscous coupled inertias. These are similar to *b* ii but since they are linear devices, they will work equally well at any amplitude. A version for grinding spindles uses a cylindrical slug in a bore with small clearance. The viscosity of the air as it is squeezed from side to side of the slug provides the necessary damping.

c. Tuned damped absorbers. These, both in theory and in practice, have by far the best performance of any passive damper. Their performance abilities can be seen from Fig. 8.1. They are usually at least a factor of three better than the Lanchester types of absorber. Designs based on the use of steel springs and oil damping, although commonly used in internal combustion engines, are usually impracticable in machine tools and have not been commercially successful. The use of elastomers, particularly nitrile or butyl rubber, which is readily obtained in O-ring seals, gives practicable designs where tuning is achieved by pre-compression of the rubber, to vary the geometry and hence the stiffness. A standard O-ring seal, when pre-compressed to approximately half its nominal

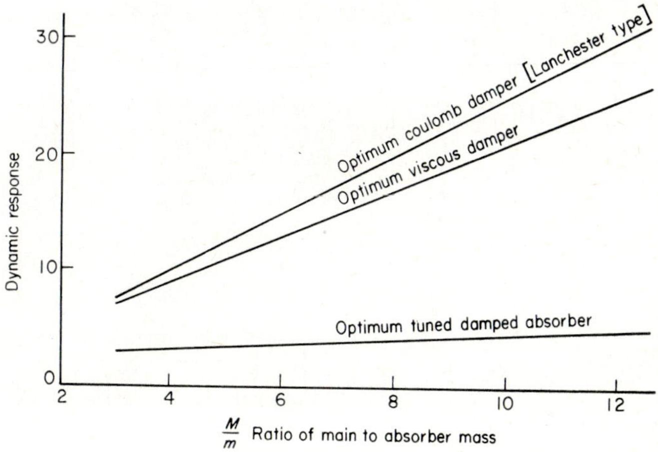

Fig. 8.1. Auxiliary damper performance.

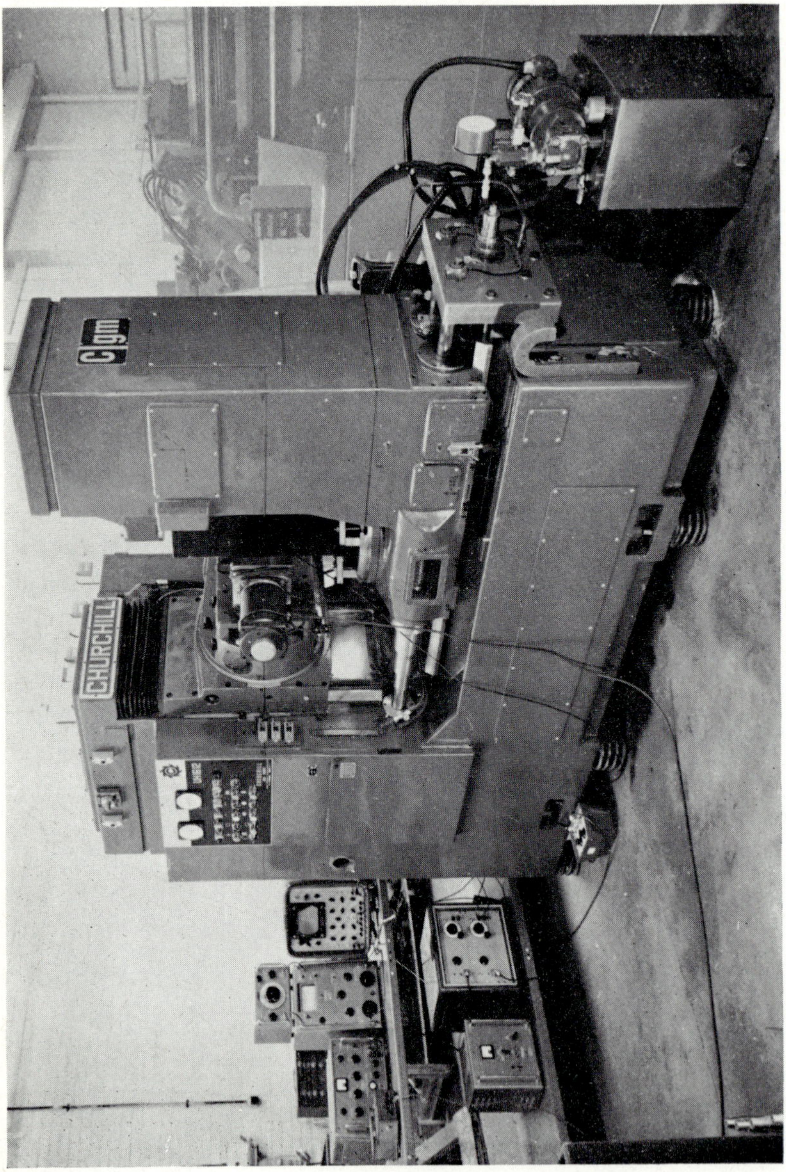

Fig. 8.2. View of hobber with tailstock auxiliary mass.

thickness, has roughly ten times the stiffness which it has at low compression. Designs in practice range from types suitable for fitting inside $\frac{7}{16}$ in diameter boring bars to designs where a mass of several tons is used to stabilize a tall column in a machine such as a post borer.

4. Criticism of the use of tuned absorbers is often based on the fact that the performance deteriorates when the system is mistuned. The performance of a mistuned damper will still normally be better than that of a Lanchester damper, particularly if the initial design is carried out with the possibility of mistuning in mind: this is discussed theoretically in Para. 7. Tuned absorbers can of course be designed to be tuned to one frequency in one direction and a different frequency at right angles allowing the simultaneous suppression of two different modes.

5. Active systems. These systems work by detecting vibrational movement of a point on a structure and using a servo system to apply a force to oppose the movement. Though these systems are very powerful, their cost, complexity and stability problems at present (1969) rule them out for all but the most special cases.

6. A typical application of a vibration damper to a hobbing machine is shown in Fig. 8.2. The dynamic response curves for the machine before fitting it and after fitting it are shown in Fig. 8.3, which gives a very good idea of the improvement available. The damper itself is the small mass at the top of the tail-stock.

7. Theory of tuned damped absorbers. Take a system as shown in Fig. 8.4 which is a schematic simplification of a linear or torsional system. In this diagram m_1, λ_1, and k_1 are parameters of the machine tool, and are therefore fixed values; whereas m_2, λ_2, and k_2 are the parameters of the vibration damper and are adjustable subject to certain practical considerations. The most obvious of these is that m_2 can generally only be some small fraction of m_1. We therefore define

$$m_2 = \mu m_1 \tag{8.1}$$

where μ is of the order of 0·1.

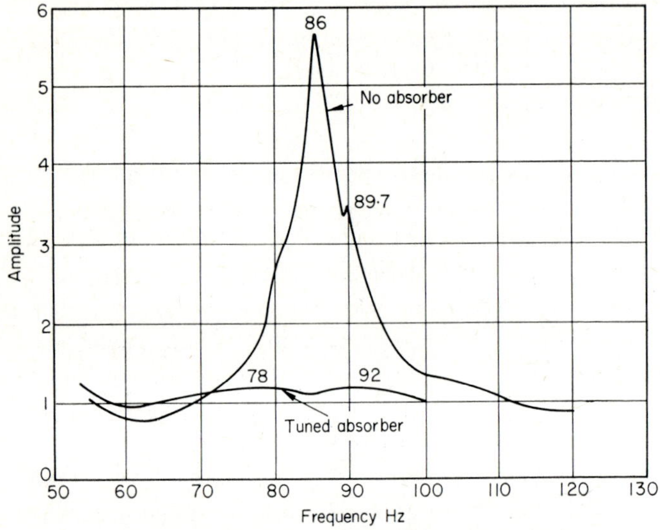

Fig. 8.3. Effect of absorber on hobber tailstock.

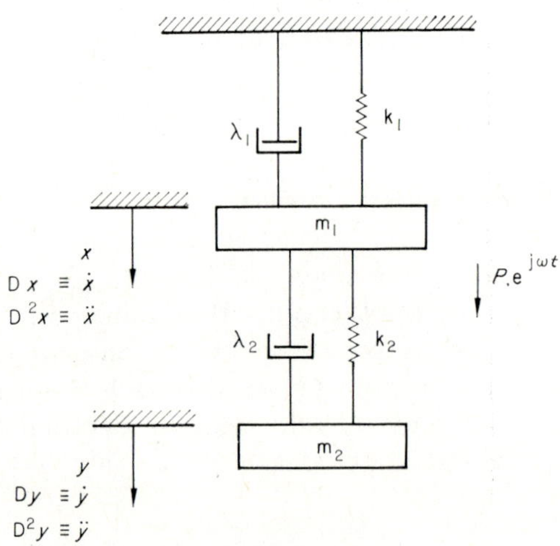

Fig. 8.4. Tuned absorber diagram.

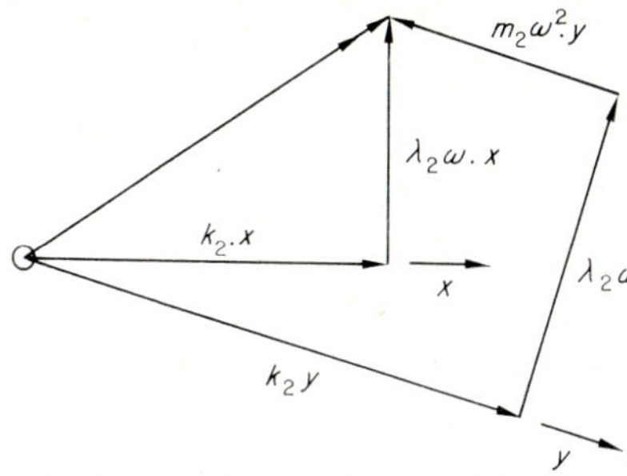

Fig. 8.5. General vector diagram for two degrees of freedom, representing Eq. 8.3.

The equation of motion for the main mass is

$$m_1\mathrm{D}^2x+\lambda_1\mathrm{D}x+k_1x = \lambda_2\mathrm{D}(y-x)+k_2(y-x)+P\,\mathrm{e}^{\mathrm{j}\omega t}$$

$$\therefore [m_1\mathrm{D}^2+(\lambda_1+\lambda_2)\mathrm{D}+(k_1+k_2)]x = [\lambda_2\mathrm{D}+k_2]y+P\,\mathrm{e}^{\mathrm{j}\omega t} \tag{8.2}$$

For the damper, or auxiliary, mass the equation of motion is

$$m_2\mathrm{D}^2y+\lambda_2\mathrm{D}(y-x)+k_2(y-x) = 0$$

$$\therefore [m_2\mathrm{D}^2+\lambda_2\mathrm{D}+k_2]y = [\lambda_2\mathrm{D}+k_2]x \tag{8.3}$$

The vector relationship for sinusoidal motion for Eq. 8.3 is shown in Fig. 8.5.

Eliminating y from Eq. 8.2 and 8.3 gives

$$[m_1\mathrm{D}^2+(\lambda_1+\lambda_2)\mathrm{D}+(k_1+k_2)]x-\frac{[\lambda_2\mathrm{D}+k_2][\lambda_2\mathrm{D}+k_2]}{[m_2\mathrm{D}^2+\lambda_2\mathrm{D}+k_2]}x = P\,\mathrm{e}^{\mathrm{j}\omega t}$$

$$\therefore [m_1m_2\mathrm{D}^4+\{m_2\lambda_1+(m_1+m_2)\lambda_2\}\mathrm{D}^3+\{\lambda_1\lambda_2+(m_1+m_2)k_2+m_2k_1\}\mathrm{D}^2 +(\lambda_2k_1+\lambda_1k_2)\mathrm{D}+k_1k_2]x = [m_2\mathrm{D}^2+\lambda_2\mathrm{D}+k_2]P\,\mathrm{e}^{\mathrm{j}\omega t} \tag{8.4}$$

This equation is represented vectorially in Fig. 8.6.
Eliminating x from Eq. 8.2 and 8.3 gives

$$[\text{As in LHS of Eq. 8.4}]y = [\lambda_2\mathrm{D}+k_2]P\,\mathrm{e}^{\mathrm{j}\omega t} \tag{8.5}$$

Writing $\omega_1^2 = \dfrac{k_1}{m_1}$, $\omega_2^2 = \dfrac{k_2}{m_2}$, $\dfrac{\lambda_1}{k_1} = \dfrac{1}{Q_1\omega_1}$ and $\dfrac{\lambda_2}{k_2} = \dfrac{1}{Q_2\omega_2}$, and dividing through by k_1k_2, Eq. 8.4 becomes

$$\left[\frac{\mathrm{D}^4}{\omega_1^2\omega_2^2}+\left(\frac{1}{\omega_2^2\omega_1Q_1}+\frac{(1+\mu)}{\omega_1^2}\cdot\frac{1}{Q_2\omega_2}\right)\mathrm{D}^3+\left(\frac{1}{\omega_1\omega_2Q_1Q_2}+\frac{(1+\mu)}{\omega_1^2}+\frac{1}{\omega_2^2}\right)\mathrm{D}^2 +\left(\frac{1}{\omega_2Q_2}+\frac{1}{\omega_1Q_1}\right)\mathrm{D}+1\right]x = \left[\frac{\mathrm{D}^2}{\omega_2^2}+\frac{1}{\omega_2Q_2}\mathrm{D}+1\right]\frac{P}{k_1}\mathrm{e}^{\mathrm{j}\omega t} \tag{8.6}$$

This equation is represented vectorially in Fig. 8.7. A family of such vector diagrams is shown in Fig. 8.12 (ii).

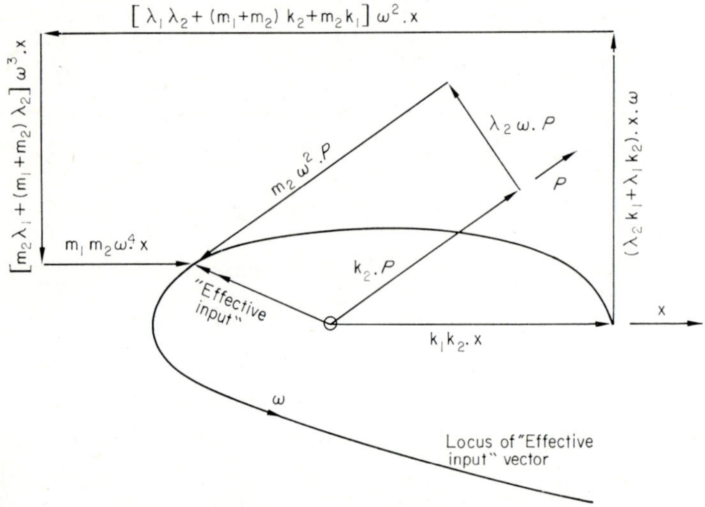

Fig. 8.6. General vector diagram for two degrees of freedom, representing Eq. 8.4.

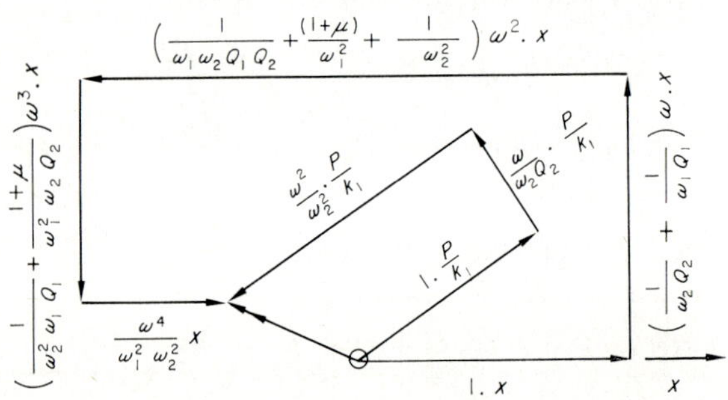

Fig. 8.7. Vector diagram in non-dimensional form.

8. An alternative way of forming the equations is to consider the relative motion between x and y: putting $(y-x) = z$.

$$[m_1\mathrm{D}^2+\lambda_1\mathrm{D}+k_1]x = [\lambda_2\mathrm{D}+k_2]z+P\,\mathrm{e}^{j\omega t} \tag{8.7}$$

and

$$[m_2\mathrm{D}^2+\lambda_2\mathrm{D}+k_2]z = -m_2\mathrm{D}^2x \tag{8.8}$$

Reforming these equations to eliminate z will give Eq. 8.4 as before: eliminating x gives

$$[m_2^2\mathrm{D}^4+m_2\lambda_2\mathrm{D}^3+(m_2k_2+\lambda_2^2)\mathrm{D}^2+2\lambda_2k_2\mathrm{D}+k_2^2]z = m_2\mathrm{D}^2P\,\mathrm{e}^{\mathrm{j}\omega t}$$

or

$$\frac{\omega_2^2}{\mathrm{D}^2}\left[\frac{\mathrm{D}^4}{\omega_2^4}+\frac{1}{\omega_2^3Q_2}\mathrm{D}^3+\left(\frac{1}{\omega_2^2}+\frac{1}{\omega_2^2Q_2^2}\right)\mathrm{D}^2+\frac{2}{\omega_2Q_2}\mathrm{D}+1\right]z = \frac{P}{k_2}\,\mathrm{e}^{\mathrm{j}\omega t} \tag{8.9}$$

Each of these methods of forming the equations throws different light on the phase relationships of the parts of the system.

9. The analysis of these equations by the usual control theory techniques is not straightforward, since the roots of quartics are not readily found in algebraic terms: Bode diagram techniques and root locus methods are therefore not helpful. The general shape of the locus in Fig. 8.6, and consideration of the family of vector diagrams that go with it, shows that in general two resonances can occur. The most helpful approach to the problem is by using a computer fitted with a plotter to obtain families of curves and vector diagrams such as appear in this chapter.*

Fig. 8.8 (i) shows a dimensionless plot of the amplitude for the main mass, m_1, as the frequency and damping ratio of the auxiliary mass, m_2, are varied. Points of importance are:

a. for no damping, the motion of the main mass may be made zero at one frequency, at the cost of two large resonances on either side of this frequency:

b. for high damping on the auxiliary mass, the two masses are effectively locked together, and the system behaves almost as a single degree of freedom system:

c. all the curves pass through two common points, marked *A* and *B* on the figure.

Fig. 8.8 (ii) shows the amplitude relationships for the relative motions of the two masses. The motion of the auxiliary mass is shown in Fig. 8.8 (iii).

Fig. 8.9 (i), (ii) and (iii) show the phase relationships for the system.

* These were obtained on an IBM 1130 computer fitted with a rotery plotter We are indebted for this program to Miss R. A. Britton of the Cambridge University Engineering Department who wrote it in Fortran IV.

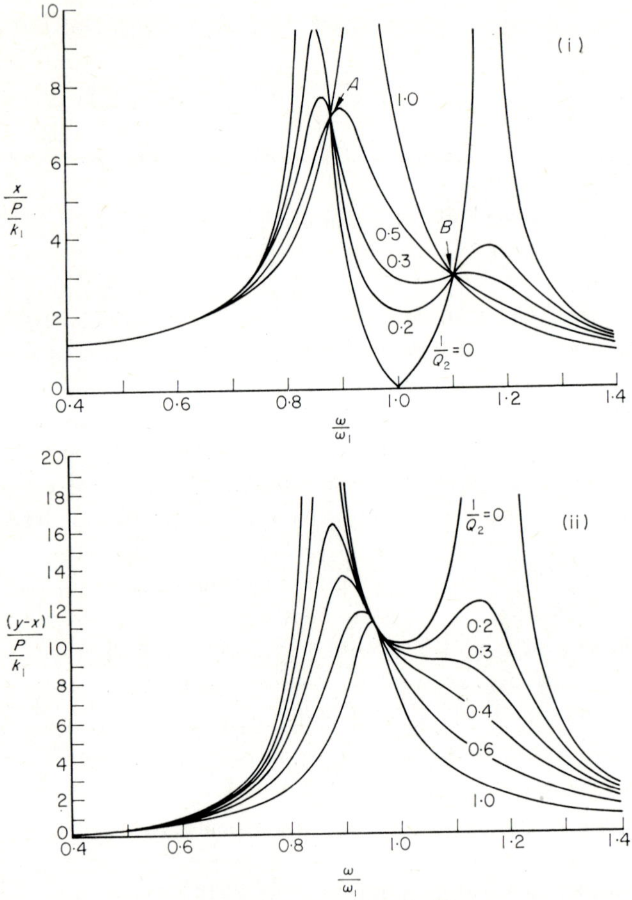

Fig. 8.8. Parts (i) and (ii).

Fig. 8.8. (i) Dynamic response of main mass $\dfrac{x}{\dfrac{p}{k_1}}$; (ii) Relative motion between masses $\dfrac{(y-x)}{\dfrac{p}{k_1}}$; (iii) Dynamic response of auxiliary mass $\dfrac{y}{\dfrac{p}{k_1}}$.

$$\frac{\omega_2}{\omega_1} = 1,\ \mu = 0{\cdot}1,\ \frac{1}{Q_1} = 0,\ \frac{1}{Q_2}\ \text{varies.}$$

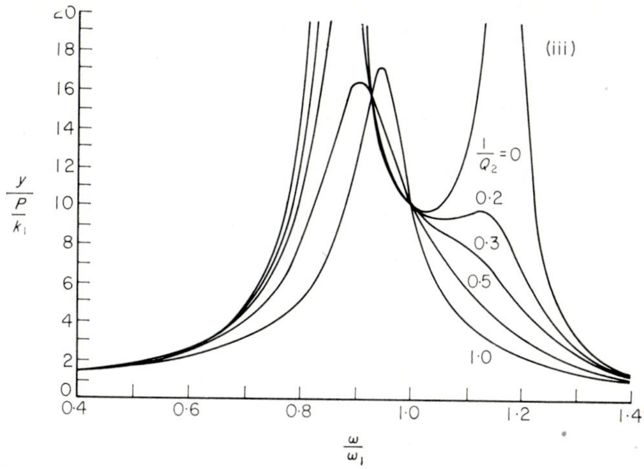

Fig. 8.8. Part (iii).

10. For situations in which a tuned vibration damper is used, Q_1 is likely to be of the order of 20. The ratio between the main mass and the mass of the damper is typically about 10, and the value of Q_2, as will be shown, is likely to be of the order of 3. The frequency ω_2 is likely, as will be shown, to be slightly less than ω_1. These facts should be kept in mind: in particular they mean that Q_1 may usually be treated as being infinite, i.e. $\lambda_1 \to 0$. The wary designer will also remember that the system here represented by m_1 and k_1 may itself be a high order system, and that the damping terms may be hysteretic. In practice neither fact need prove too worrying.

11. No damping on either system $\lambda_1 = \lambda_2 = 0$. For this case the vector diagrams collapse into a straight line, and so does the vector locus. From Fig. 8.6, with $\lambda_2 = 0$, if $m_2\omega^2 - k_2 = 0$, then $x = 0$; but note that y can have some finite value. From Eq. 8.2, when $x = 0$, $y = -\frac{P}{k_2}$. In other words, the damper mass is in anti-phase to P, and its amplitude produces a force on the main mass exactly balancing that due to P. This may be seen in Figs. 8.8 (i), (ii) and (iii). The damper is therefore completely effective for $\omega^2 = \frac{k_2}{m_2}$.

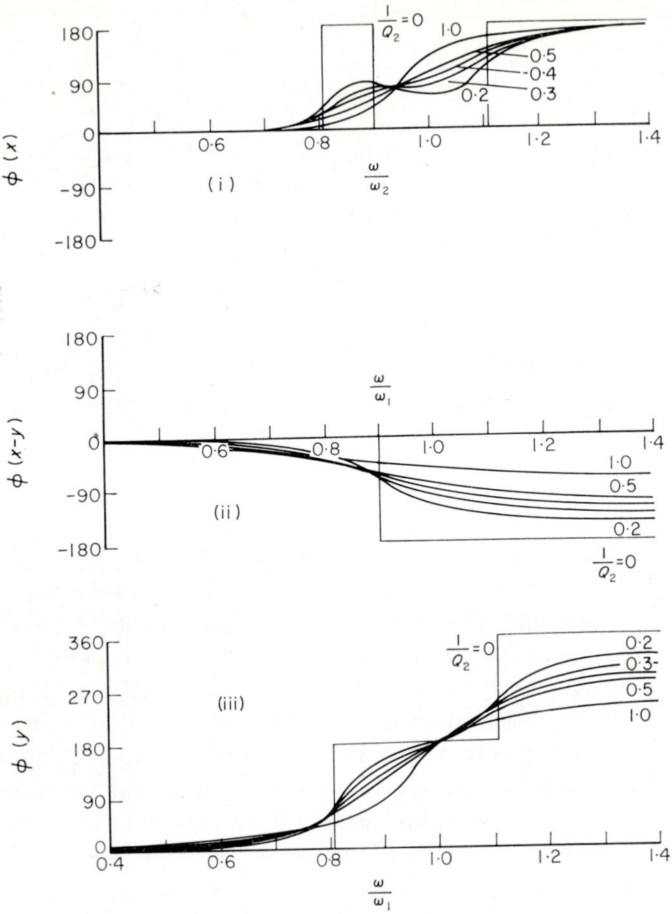

Fig. 8.9. Phase relationship between applied force and masses. $\frac{\omega_2}{\omega_1} = 1, \mu = 0{\cdot}1,$

$\frac{1}{Q_1} = 0, \frac{1}{Q_2}$ varies.

Resonance of m_1 will occur when the input can be zero for finite x. This occurs when

$$\frac{\omega^4}{\omega_1^2\omega_2^2} - \left[\frac{1+\mu}{\omega_1^2} + \frac{1}{\omega_2^2}\right]\omega^2 + 1 = 0 \tag{8.10}$$

which factorizes as a quadratic in ω^2 to give two values, one above and one below ω_1. This may be seen in Fig. 8.8 (i). These are the frequencies of the two normal modes of the system.

12. Amplitude of main mass independent of λ_2. Fig. 8.10 shows the amplitude, Fig. 8.11 the phase and Fig. 8.12 the vector relationships for a system so arranged that $\frac{\omega_2}{\omega_1} = \frac{1}{1+\mu}$. The reason for choosing this value will shortly appear. The families of curves shown in Figs. 8.8 (i) and 8.10 (i) drawn by a computer for $\frac{1}{Q_1} = 0$ show that independent of the value of Q_2 they all go through two points A and B. If these two points can be made to

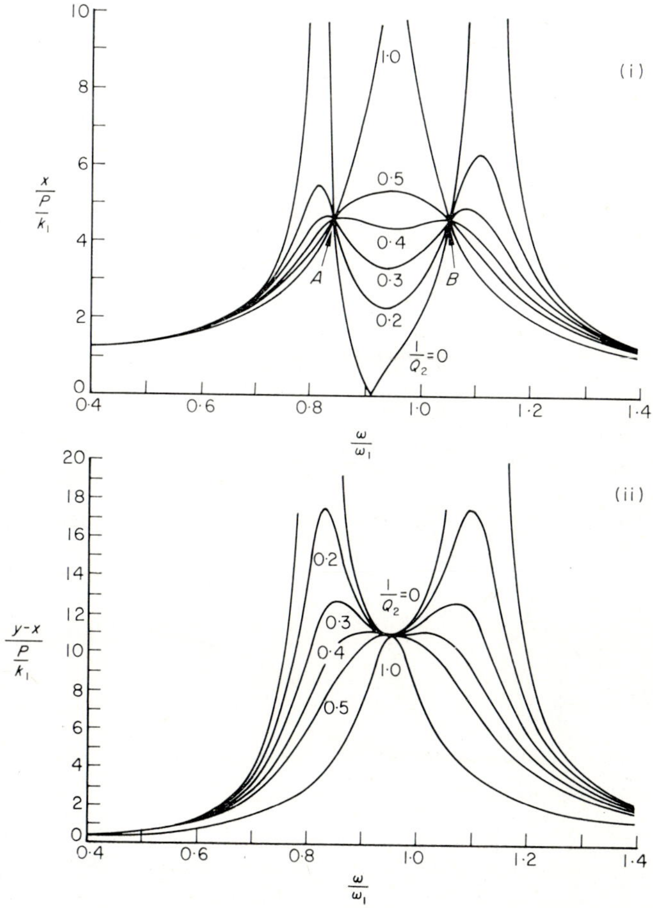

Fig. 8.10. Dynamic response curves for optimum frequency ratio of main and auxiliary systems. $\frac{\omega_2}{\omega_1} = \frac{1}{1+\mu}$, $\mu = 0{\cdot}1$, $\frac{1}{Q_1} = 0$, $\frac{1}{Q_2}$ varies.

have equal amplitudes, as in Fig. 8.10 (i), a result very close to the optimum can be obtained if a flat resonance curve is required. The conditions for this to occur can be seen clearly from the vector diagrams Fig. 8.13 which are redrawn, *not to scale*, from the computer-drawn diagrams of Fig. 8.12.

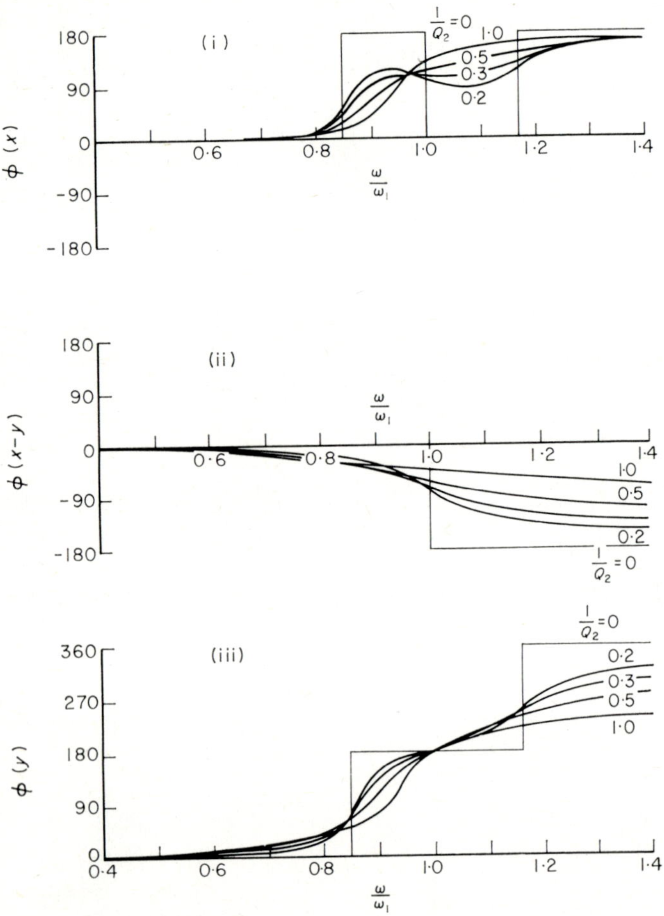

Fig. 8.11. Phase relationship between applied force and masses. $\dfrac{\omega_2}{\omega_1} = \dfrac{1}{1+\mu}$,

$$\mu = 0.1, \frac{1}{Q_1} = 0, \frac{1}{Q_2} \text{ varies.}$$

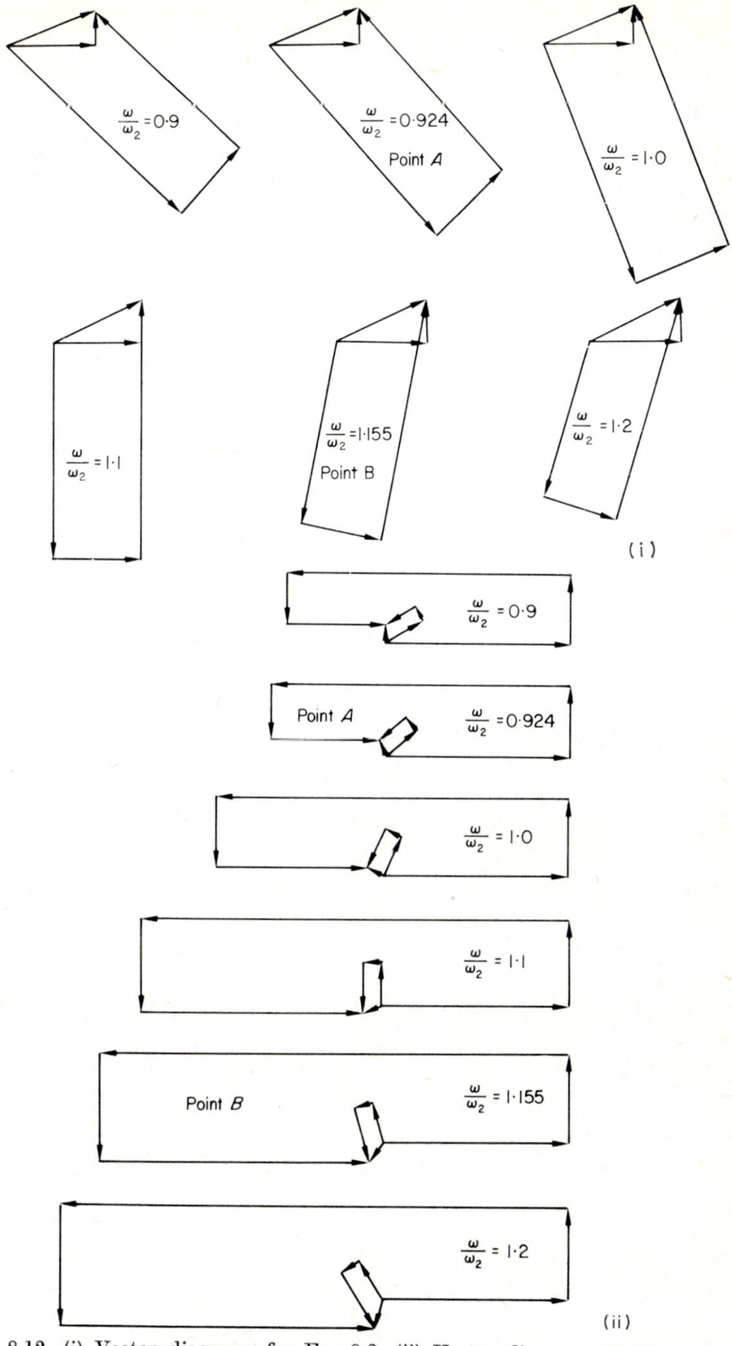

Fig. 8.12. (i) Vector diagrams for Eq. 8.2. (ii) Vector diagrams for Eq. 8.3. $\frac{\omega_2}{\omega_1} = \frac{1}{1+\mu}$, $\mu = 0{\cdot}1$, $\frac{1}{Q_1} = 0$, $\frac{1}{Q_2} = 0{\cdot}4$.

Considering the two triangles in Fig. 8.13 each has only one side dependent on λ_2. If therefore these triangles are congruent, the relationship between P and x is independent of λ_2.

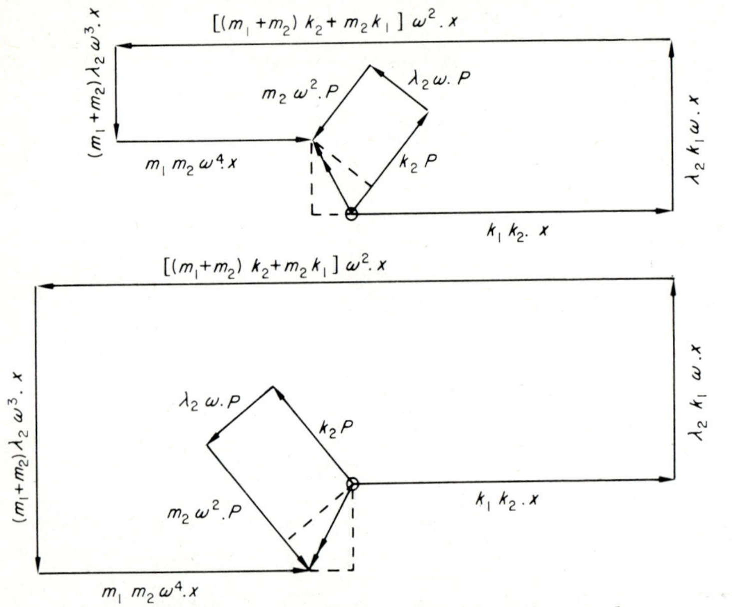

Fig. 8.13. Vector diagrams for fixed points. *Not to scale.*

The condition for this to be so is

$$\frac{[(m_1+m_2)\lambda_2\omega^3-\lambda_2k_1\omega]x}{\lambda_2\omega\,.\,P}=\frac{[-\{(m_1+m_2)k_2+m_2k_1\}\,\omega^2+m_1m_2\omega^4+k_1k_2]x}{(k_2-m_2\omega^2)P} \tag{8.11}$$

(Note that the signs of the vectors must be correctly preserved.)

Putting $m_2=\mu m_1$; $\omega_1^2=\dfrac{k_1}{m_1}$; $\omega_2^2=\dfrac{k_2}{m_2}$;

$$\left[(1+\mu)\frac{\omega^2}{\omega_1^2}-1\right]\left[1-\frac{\omega^2}{\omega_2^2}\right]k_1k_2=k_1k_2\left[-\left(\frac{1+\mu}{\omega_1^2}+\frac{1}{\omega_2^2}\right)\omega^2+\frac{\omega^4}{\omega_1^2\omega_2^2}+1\right]$$

$$\therefore\ \frac{(2+\mu)}{\omega_1^2\omega_2^2}\,\omega^4-\left[\frac{2(1+\mu)}{\omega_1^2}+\frac{2}{\omega_2^2}\right]\omega^2+2=0$$

$$\therefore\ \omega^4-\left(\frac{2}{2+\mu}\right)[(1+\mu)\omega_2^2+\omega_1^2]\omega^2+\frac{2}{2+\mu}\,\omega_1^2\omega_2^2=0 \tag{8.12}$$

Call the two roots of this equation ω_a^2 and ω_b^2, and the amplitude of the displacements for these values x_a and x_b. Then from the vector diagrams in Fig. 8.13

$$\lambda_2\omega_a P = -(m_1+m_2)\lambda_2\omega_a^3 x_a+\lambda_2 k_1\omega_a x_a$$

and

$$\lambda_2\omega_b P = +(m_1+m_2)\lambda_2\omega_b^3 x_b-\lambda_2 k_1\omega_b x_b$$

$$\therefore \frac{\frac{P}{k_1}}{x_a} = -\frac{(1+\mu)}{\omega_1^2}\,\omega_a^2+1 \tag{8.13}$$

and

$$\frac{\frac{P}{k_1}}{x_b} = +\frac{(1+\mu)}{\omega_1^2}\,\omega_b^2-1 \tag{8.14}$$

When $x_a = x_b$, subtracting gives

$$\omega_a^2+\omega_b^2 = \frac{2}{1+\mu}\,\omega_1^2 \tag{8.15}$$

From Eq. 8.12,

$$\omega_a^2+\omega_b^2 = \frac{2}{2+\mu}\,(1+\mu)\omega_2^2+\omega_1^2 \tag{8.16}$$

Equating Eq. 8.15 and Eq. 8.16 gives

$$\frac{\omega_2}{\omega_1} = \frac{1}{1+\mu} \tag{8.17}$$

(*Caution*: This is *not* the same result as obtains for Fig. 8.15 and Equation 8.23.)

The values for A and B are independent of λ_2 (or Q_2). They are *not* independent of λ_1 (or Q_1), and its effect is shown in Fig. 8.14 (i) and (ii).

The value of $x = x_a = x_b$ at A and B may be obtained by adding Eq. 8.13 and 8.14 and substituting from Eq. 8.17. Then using Eq. 8.12

$$2\frac{\frac{P}{k_1}}{x} = \frac{(1+\mu)}{\omega_1^2}\,(-\omega_a^2+\omega_b^2) = \frac{(1+\mu)}{\omega_1^2}\sqrt{\frac{4\omega_1^4}{(1+\mu)^2}-\frac{4.2\omega_1^4}{(1+\mu)^2(2+\mu)}}$$

$$\therefore \frac{x}{\frac{P}{k_1}} = \sqrt{\frac{2+\mu}{\mu}} \tag{8.18}$$

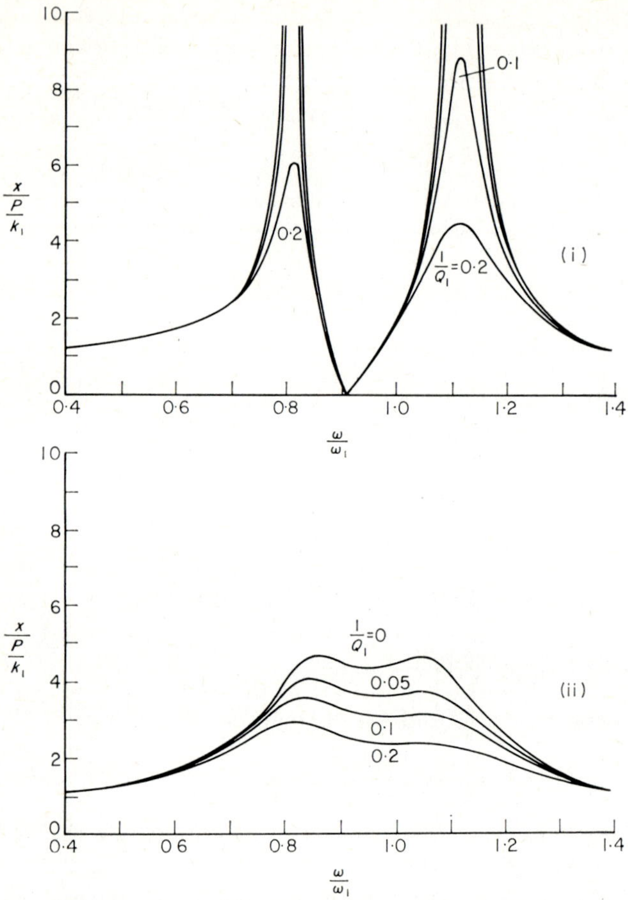

Fig. 8.14. Effect of variation of damping on main mass. $\frac{\omega_2}{\omega_1} = \frac{1}{1+\mu}$, $\mu = 0{\cdot}1$, $\frac{1}{Q_1}$ varies. (i): $\frac{1}{Q_2} = 0$; (ii): $\frac{1}{Q_2} = 0{\cdot}4$.

The 'best', i.e. the flattest topped resonance curve, is sometimes taken as being when the curve passes horizontally through A or B. The values for this can be shown to be that

$$Q_2 = \sqrt{\frac{2(1+\mu)^3}{\mu\left(3 \pm \sqrt{\frac{\mu}{\mu+2}}\right)}} \tag{8.19}$$

The average value is sometimes taken, giving

$$Q_2 = \sqrt{\frac{2(1+\mu)^3}{3\mu}} \tag{8.20}$$

Investigation with a computer shows that this value is sometimes significantly different from the optimum.

13. Effect of damping on main mass. If a damper is required, the damping on the main mass is slight and Q factors between 10 and 20 are usual. Fig. 8.14 (i) and (ii) show the effect of varying Q_1. A special case, discussed in the next paragraph, is of interest if the damping on the main mass is unknown or variable.

14. Effective input opposed to motion of main mass. Fig. 8.15 shows a special case, not to scale, of the family of vector diagrams shown in Fig. 8.16 and Fig. 8.17 where the effective input to the main mass is exactly opposed to the motion of the mass, and when P is in phase with its velocity. Equating opposite vertical (parallel to the imaginary axis) sides of the two vector diagrams

$$\frac{\lambda_2 k_1 + \lambda_1 k_2}{m_2\lambda_1 + (m_1+m_2)\lambda_2} = \omega^2 = \frac{k_2}{m_2} \tag{8.21}$$

Cross multiplying and simplifying yields

$$\frac{m_1+m_2}{k_1} = \frac{m_2}{k_2} \tag{8.22}$$

which is independent of both λ_1 and of λ_2.

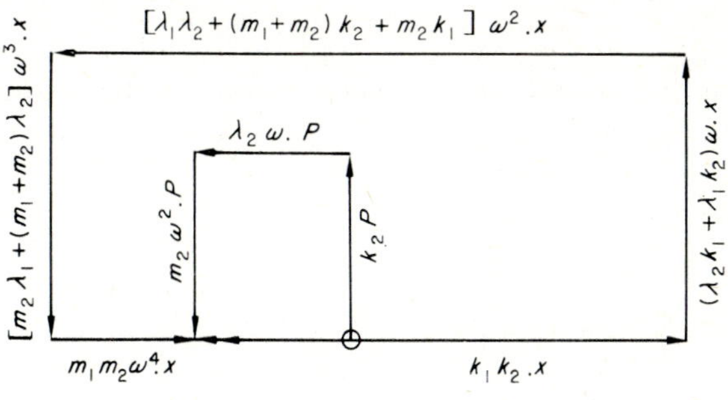

Fig. 8.15. Vector diagram, 90° phase. *Not to scale.*

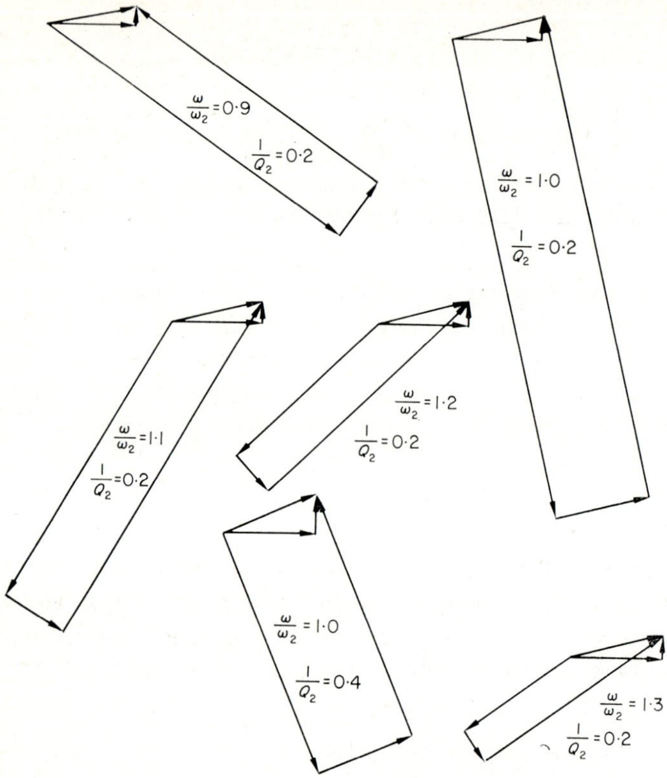

Fig. 8.16. Vector diagrams for $\frac{\omega_2}{\omega_1}=\sqrt{\frac{1}{1+\mu}}$, $\mu=0{\cdot}1$, $\frac{1}{Q_1}=0$.

Writing $m_2=\mu m_1$

$$\frac{\omega_2}{\omega_1}=\sqrt{\frac{1}{1+\mu}} \tag{8.23}$$

This is *not* the same result as in Eq. 8.17.

Typically, μ is $\frac{1}{10}$, so ω_2 is slightly less than ω_1.

Equating horizontal, or real, sides of the vector diagram gives

$$\lambda_2\omega\,.\,P=[\{\lambda_1\lambda_2+(m_1+m_2)k_2+m_2k_1\}\omega^2-m_1m_2\omega^4-k_1k_2]x$$

But

$$\omega^2=\frac{k_2}{m_2}=\frac{k_1}{m_1+m_2}$$

$$\therefore\ \lambda_2\sqrt{\frac{k_2}{m_2}}\,\frac{P}{x}=\lambda_1\lambda_2\omega_2^2+k_2^2 \quad\text{or}\quad \frac{P}{x}=\lambda_1\omega_2+\frac{k_2^2}{\lambda_2\omega_2}$$

$$\text{and}\quad \frac{x}{P}=\frac{1}{\lambda_1\omega_2+Q_2k_2}\quad\text{or}\quad \frac{x}{\dfrac{P}{k_1}}=\frac{1}{\dfrac{\omega_2}{\omega_1Q_1}+\dfrac{\mu Q_2}{1+\mu}} \tag{8.24}$$

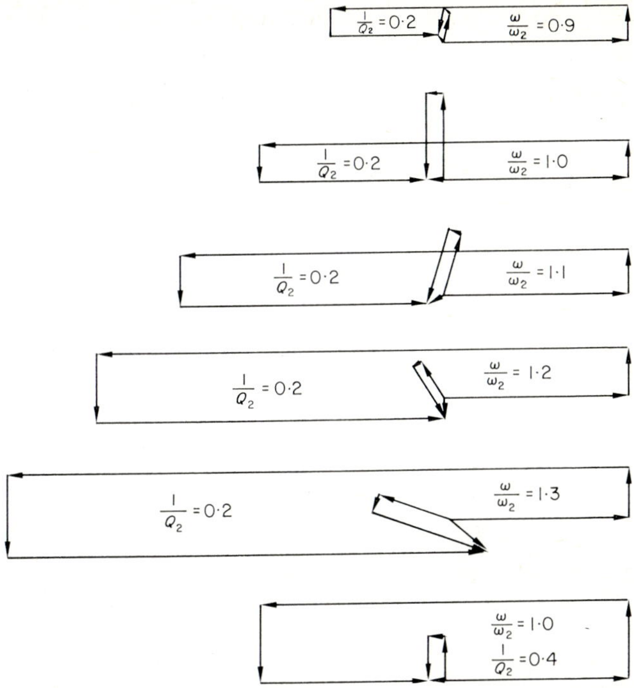

Fig. 8.17. Vector diagrams for $\frac{\omega_2}{\omega_1} = \sqrt{\frac{1}{1+\mu}}$, $\mu = 0{\cdot}1$, $\frac{1}{Q_1} = 0$

15. Untuned damper. Torsional vibration dampers are often used in which only damping is present, and no spring exists; i.e. $k_2 = 0$. Modern viscostatic silicone fluids make these easy to design and reliable. They obviate the difficulties inherent in the design of tuned dampers, where the main resonant frequency is often variable (due to backlash in gear trains); and also the difficulty of designing a damper spring to allow adequate amplitude of vibration. While they are not as effective as tuned dampers, they are sometimes useful, particularly for a system with a number of resonant frequencies. Fig. 8.18 (i) shows families of response curves for varying amounts of damping on the auxiliary mass. Fig. 8.18 (ii) shows the effect of damping on the main system.

These curves have in fact not been plotted for an entirely untuned damper, since the programme, when written in non-dimensional form, would not handle this special case. Making $\frac{\omega_2}{\omega_1} = \frac{1}{100}$ gives a very close approximation.

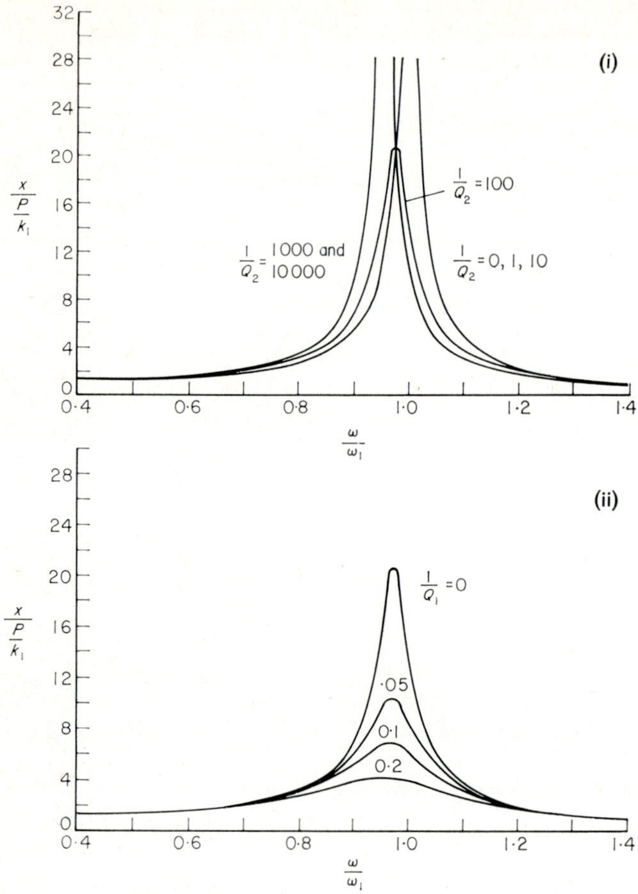

Fig. 8.18. Response curves for viscous damper $\left(\frac{\omega_2}{\omega_1} = \frac{1}{100}\right)$.

(i) $\mu = 0{\cdot}1; \frac{1}{Q_1} = 0; \frac{1}{Q_2}$ varies. (ii) $\mu = 0{\cdot}1$; Q_1 varies; $\frac{1}{Q_2} = 100$.

16. Practical design considerations. No general theory exists for the design of dampers for multi-degree of freedom systems, and for optimum results each must be examined with the aid of an analogue or digital computer. A result very close to the optimum will always be obtained with a damper tuned as in Para. 12 for the lowest resonant frequency.

In practice, no two machines are identical, and provision should be made for tuning dampers on the test bed. This may conveniently be done as discussed in Para. 8.3*c* above. The damping properties of both nitrile and butyl rubbers approximate more closely to hysteretic than to viscous damping, but this has little practical effect on the theory developed above, as may be shown by formal analysis. In particular, the result of Eq. 8.17 is unaffected. Fig. 8.3 shows the sort of result obtainable with care.

17. Problems

1. Why should provision always be made for adjusting tuned vibration dampers on test? Would this in general be expected to be more important for a machine with a cast or a fabricated frame?

2. A large (e.g. 12 ft table) gear hobbing machine or a large vertical borer is limited in its production capacity by chatter. What proposals would you make for investigating the problem with a minimum loss of production time, and what proposals would you make for improving the machine at minimum cost? (See also Chap. 10.)

9. Miscellaneous components

1. Dead stops. When accurate positioning is required, slides are commonly moved up to dead stops. The contact forces produced by this can be of very large magnitudes even at low velocities. The full analysis of the resulting stress waves is extremely complex, and must be made along the lines of Chap. 7.15. An elementary approach which gives the exact analysis for these forces for one special case, and the order of magnitude for more general cases, is given below.

2. Suppose that a cylindrical steel rod moving with velocity v hits an object which stops the end instantaneously. Since the rod is elastic the remainder of the rod will continue to move forward, and to compress itself until the whole rod has come to rest. The free end will then start to move back again. The process is perhaps easier to visualize if the rod is thought of as being made of rubber.

At a short time after the impact when only part of the rod has come to rest, Fig. 9.1, the strain energy in the compressed portion must equal the kinetic energy that has been destroyed, assuming no losses. If p is the rise in pressure in the rod, E the Young's Modulus and ρ the density, A the cross-sectional area, and the length of the rod at rest is l,

$$\text{Kinetic Energy Lost} = \frac{1}{2g}\rho l A v^2$$

$$= \text{Strain Energy} = \frac{1}{2} p A \delta l$$

$$\text{But } p = E\frac{\delta l}{l}$$

$$\therefore p = v\sqrt{\frac{E\rho}{g}} \qquad (9.1)$$

The pressure rise if v *is in in/s* will be

$$p = v\sqrt{\frac{30\times 10^6 \times 0{\cdot}283}{32{\cdot}2\times 12}} = 148v \text{ lbf/in}^2$$

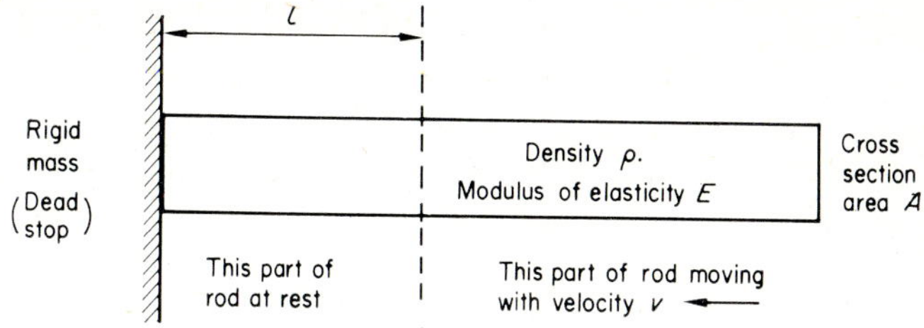

Fig. 9.1. Impact of rod.

A velocity of just over 15 in/s will therefore give rise to a pressure of 1 ton/in^2. Since the yield point for mild steel is about 12 ton/in^2, an impact of about 15 ft/s or 10 m.p.h. will be enough to cause plastic deformation of the steel. This is why many cams and dead-stops come to be deformed. A velocity of 15 ft/s is acquired in a free fall under gravity of about 3·5 ft.

The story is not quite as simple as this: the object with which the collision is taking place is itself not rigid, and this cushions the impact slightly. The order of magnitude, however, is correct.

The rod will tend to bounce off the stop when the stress wave has travelled to the end and back again. The implication for dead stops is that if approached fast they will not give consistent results, since the part to be positioned will tend to bounce off the stop by some indeterminate amount.

The velocity of propagation of a pressure wave along a rod is

$$V = \sqrt{\frac{Eg}{\rho}} \tag{9.2}$$

For steel putting $E = 30 \times 10^6$ psi, $\rho = 0{\cdot}283$ lb/in^3,

$$V = \sqrt{\frac{30 \times 10^6 \times 32{\cdot}2 \times 12}{0{\cdot}283}} = 16\ 900 \text{ ft/s}.$$

For cast iron the value would be about 11 000 ft/s.

3. Cams. Cams are usually the easiest method of producing any desired motion of an irregular variety. Often four-bar linkages would produce kinematically better solutions, but their analysis tends to be difficult, and is little understood. While cams are treated extensively in most textbooks on mechanics of machines the treatment rarely emphasizes the fundamental physica principles involved, and as a result machine tools commonly contain cams which have been designed to do the impossible.

The displacement most commonly desired is the square wave train shown in Fig. 9.2 (i) at the top of the column. This is impossible to achieve, since to do so would require an infinite velocity and a doubly infinite acceleration, with infinite forces as a consequence.

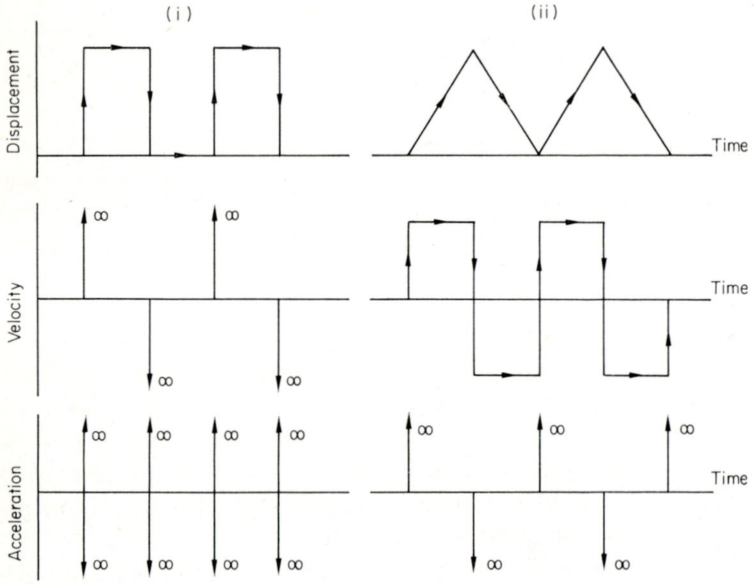

Fig. 9.2. Cam displacements, velocities and accelerations.

Another displacement which is often desired is that shown in Fig. 9.2 (ii) where the velocities are finite but the acceleration is infinite, and again the forces required are infinite.

4. The lowest order curve to give finite accelerations, and hence finite forces, is the parabola, but this gives rise to an abrupt change in the acceleration, or 'jerk', which is a step function, and which will give rise to unwanted stress waves in the system. For this reason, the lowest order curve which should normally be used is a cubic. Railway engineers have for many years used transition curves which give uniform rates of change of acceleration, so that when entering a curve the trains and their passengers are not subjected to jerks. These curves are, however, relatively complicated to handle, and the best compromise curve for most purposes is a sinusoidal one. This has the advantage both of being familiar, and thoroughly tabulated in accessible places. Both transition and sinusoidal curves approximate to cubics.

5. The considerations above lead directly to the conclusion that harmonic cams, consisting of arcs of circles tangential to straight lines or other circles, must be avoided in fast running machinery, since they theoretically involve infinite rate of change of accelerations. These clearly cannot occur, but very large ones do; in one machine tool, changing from a harmonic cam to a cam with sinusoidal changes of curvature resulted in the accelerations at the end of a long shaft, which it was driving, being reduced by a factor of 10, and the machine running noticeably more smoothly. Cams must therefore be designed with the acceleration diagram kept firmly in mind.

6. Bad cam design frequently occurs in scroll lift cams, when the follower is allowed to fall back under the spring load. This leads to very high jerks on the machine, and to the hammering of the cam which can so often be seen. The follower bounces, and finally only reaches the desired steady position very much more slowly than it would have done had the follower been guided by a suitably designed curve.

7. The basis of the design of sinusoidal transition curves is as follows:

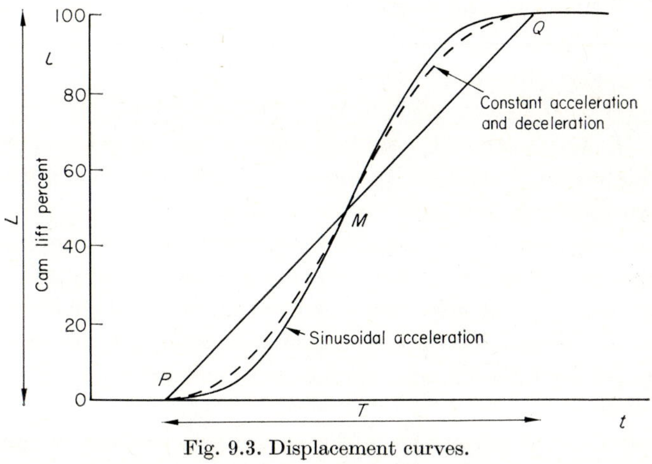

Fig. 9.3. Displacement curves.

Suppose that in Fig. 9.3 the lift needs to be altered from the value at P to the value at Q in time T, and that the acceleration is to be sinusoidal. We can write

$$\text{Acceleration} = \frac{d^2l}{dt^2} = A \sin\frac{2\pi t}{T} \tag{9.3}$$

$$\therefore \text{Velocity} = \frac{dl}{dt} = -A\frac{T}{2\pi}\cos\frac{2\pi}{T}t + B, \text{ where } B \text{ is a constant.}$$

$$\therefore \text{Displacement} = l = \frac{-AT^2}{2\pi}\sin\frac{2\pi t}{T} + Bt + C, \text{ where } C \text{ is a}$$

constant.

Initial conditions are that $t = 0, \quad l = 0, \quad \frac{dl}{dt} = 0$

Final conditions are that $t = T, \quad l = L, \quad \frac{dl}{dt} = 0$

$$\left.\begin{aligned} \therefore 0 &= -A\frac{2\pi}{T} + B \\ 0 &= C \\ L &= BT \end{aligned}\right\} \qquad \begin{aligned} \therefore A &= \frac{2\pi L}{T^2} \\ B &= \frac{L}{T} \\ C &= 0 \end{aligned}$$

$$\therefore \text{Acceleration} = \frac{2\pi L}{T^2}\sin\frac{2\pi}{T}t \tag{9.4}$$

$$\text{Velocity} = -\frac{L}{T}\cos\frac{2\pi}{T}t + \frac{L}{T} \tag{9.5}$$

$$\text{Displacement} = -\frac{L}{2\pi}\sin\frac{2\pi}{T}t + \frac{L}{T}t \tag{9.6}$$

(N.B. If, as in cams for relieving lathes, the initial velocity is appreciable, the initial conditions will be different, and so will the constants.)

These equations show that the lift curve is represented by a straight line between P and Q, on to which is superimposed a sine wave. The resultant curve is shown in Fig. 9.3, and a table of the values is given in Fig. 9.4. For comparison the lift curve is also shown in Fig. 9.3 for a cam designed for constant acceleration and deceleration, which gives the minimum accelerations possible, but suffers from infinite jerks. These are very evident as hammer blows on the machine, which inevitably cause cam bounce. The relationship between these two acceleration curves is shown in Fig. 9.5.

Fig. 9.4.

Percentage of transition time	Percentage lift for const. acceleration	Percentage lift for sine acceleration
0	0·00	0·00
5	0·50	0·08
10	2·00	0·64
15	4·50	2·11
20	8·00	4·85
25	12·50	9·07
30	18·00	14·85
35	24·50	22·11
40	32·00	30·64
45	40·50	40·08
50	50·00	50·00
55	59·50	59·92
60	68·00	69·36
65	75·50	77·89
70	82·00	85·15
75	87·50	90·93
80	92·00	95·15
85	95·50	97·89
90	98·00	99·36
95	99·50	99·92
100	100·00	100·00

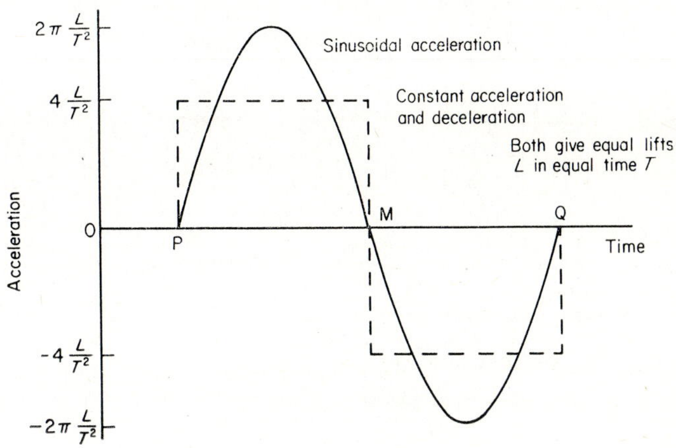

Fig. 9.5. Cam accelerations.

8. In practice the difference between the two lift curves may be a matter of only a few thousandths of an inch, but such differences can in practice make very large differences in performance. This analysis is however still incomplete, since it does not take account of oil films between cam and follower, nor of Hertzian deformation of them due to the contact stresses.

9. Roller followers should be avoided in fast running mechanisms, since high angular accelerations of the roller become theoretically necessary. In practice, adequate tangential forces often do not exist to provide these, and as a result serious rubbing occurs between roller and cam due to break down of the oil film.

10. Translation versus rotation of a mass. A problem which sometimes occurs in reciprocating motion is whether a mass should be moved bodily to and fro in slides, or whether it is better rotated about a pivot. The problem is shown schematically in Fig. 9.6 (i) and (ii). In each case the LH bottom corner of the block is to move through $b \, . \sin \omega t$. Resolving and taking moments for the two cases gives:

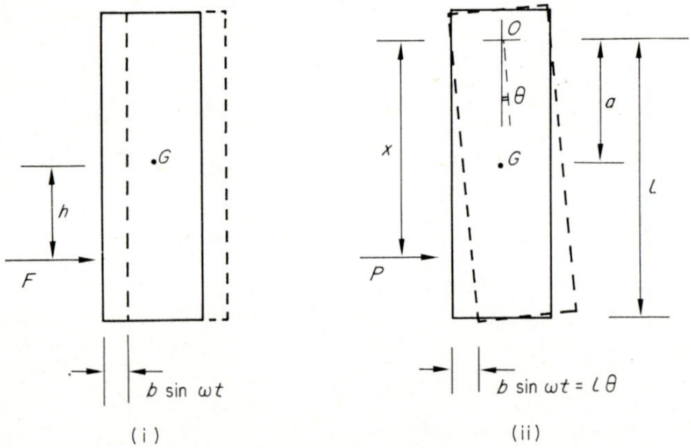

Fig. 9.6. Mass translation and rotation.

In Fig. 9.6 (i), Horizontal Force $= F = mb\omega^2 \sin \omega t$ (9.7)

and Couple on Guides $= Fh$ (9.8)

In Fig. 9.6 (ii), $m(a^2+k^2)\mathrm{D}^2\theta = Px$

$$\therefore P = m\frac{(a^2+k^2)}{x} \cdot \frac{b}{l}\,\omega^2 \sin \omega t \qquad (9.9)$$

Note that $P = \frac{(a^2+k^2)}{lx} \cdot F$ (9.10)

and in general a reaction occurs at the pivot.

Consider two special cases:

a. If P is applied at Centre of Percussion, *so that no reaction occurs at O*,

$$x = \frac{a^2+k^2}{a} \qquad (9.11)$$

and

$$P = \frac{a}{l} \cdot F \qquad (9.12)$$

b. If the mass is pivoted about G, i.e. O and G coincide, $a = 0$.

No centre of percussion exists and $P = \frac{k^2}{lx} \cdot F$ (9.13)

Now l must be $> k$ and h is probably $> k$, so this gives a reduction in P compared with F, but a horizontal reaction exists at the pivot

$$R = \left[1 - \frac{ax}{a^2+k^2}\right]P; \qquad (9.14)$$

this will probably be small.

No clear-cut answer can be given to the problem as posed, but the method of analysis for any given case is straightforward.

11. Problem

1. If you are mathematically minded, compare the use of (*a*) transition curves used by railway engineers discussed in mathematical texts in intrinsic co-ordinates, (*b*) power series, with the sinusoidal transition curves recommended in this chapter.

10. Dynamic measurements

1. Force Measurements. Measurement of cutting forces requires dynamometers which can measure forces up to the order of a ton at frequencies up to about 500 Hz. The components of force in three orthogonal axes will usually be required, and in the case of operations such as drilling or gear-cutting, torque must also be measured. To ensure that the dynamometer characteristics do not interfere with the cutting process, the dynamometer should be at least ten times as stiff as the machine tool on which it is being used and should have a natural frequency well above the operating range.

2. Strain gauges can sometimes be put on to existing parts of a machine, such as centres, to give information, but the geometry is not usually satisfactory. The most common form of machine-tool dynamometer is the octagonal ring type developed by MIT from the normal proving ring. A typical design is shown in Fig. 10.1. On each measuring element the strain gauges on the vertical faces measure vertical force while those on the 45° faces register longitudinal force.

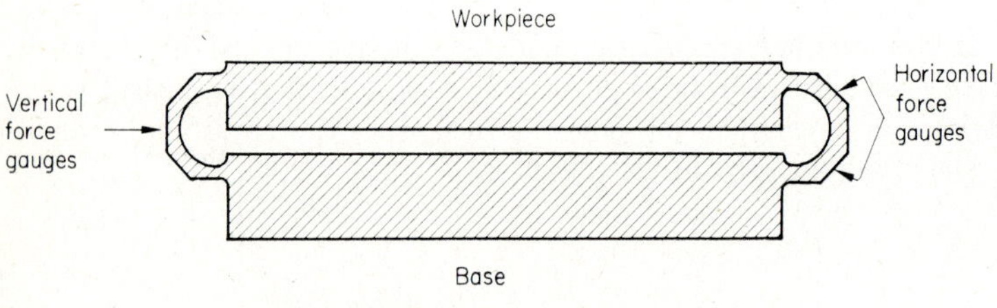

Fig. 10.1. Section of octagonal ring dynamometer.

This type of dynamometer will measure two forces and two torques but will not measure the third force, and is expensive to make since it must be milled from the solid to give a satisfactory performance.

3. An extremely versatile design of dynamometer which has been developed at the Cambridge University Engineering Laboratory from a suggestion by one of the authors (DBW), Fig. 10.2, uses a simple turned member of 'top-hat' section. The measuring section is a thin cylindrical shell attached to rigid top and bottom plates to ensure stress equalization.

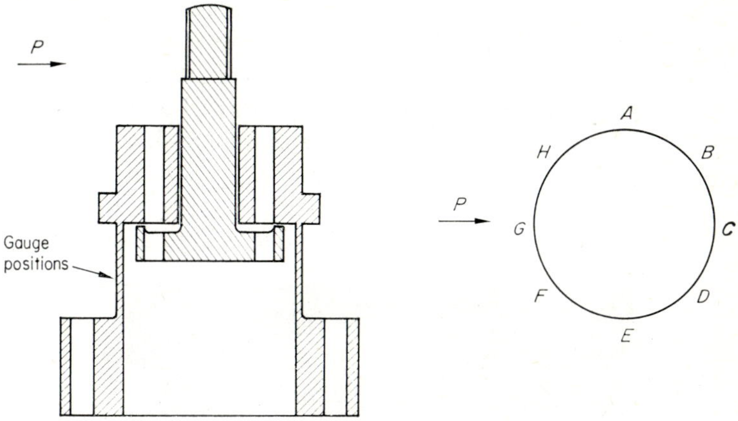

Fig. 10.2. Section of top-hat dynamometer.

The vertical force is measured regardless of position by summing the outputs from vertical strain gauges at B, D, F and H. Transverse forces or couples do not affect the reading, since they produce equal and opposite strains on opposite sides of the dynamometer, e.g. at B and F; since the strains are equal and opposite the summation cancels them.

A transverse force P as shown in the diagram will give shear at A and E while its moment will give vertical compression at C and tension at G. Strain gauges mounted at 45° to measure the shear at A and E will give outputs whose sum gives the force P while gauges mounted vertically at G and C will give outputs whose difference gives the moment of P about the centre of the cylindrical shell. A force in direction AE with a moment about GC can be measured correspondingly.

The remaining torque about a vertical axis is recorded by using gauges at 45° to measure shear by summing all clockwise shears in the bridge.

Since modern a.c. carrier strain gauge bridge systems can measure strain down to 1×10^{-7}, dynamometers can be made to register 1 lbf yet withstand 20 000 lbf, covering all normal machine-tool requirements. The upper limit is determined by the elastic properties of the material. A working cross-section in the dynamometer of roughly half a square inch will cover this range, although it is normally advisable to increase area to reduce dynamometer deflections. A dynamometer of this design is shown in use in Fig. 10.3 on a hobbing machine, while Fig. 10.4 shows traces of forces obtained when shaving a gear between centres using similar dynamometers. The time scale of the bottom three traces is 50 times that for the top three traces. Traces 1 and 4 show the tangential component of force, traces 2 and 5 show the axial components of force, and traces 3 and 6 show the separating components of force.

The centre distance between wheel and shaving cutter was kept

Fig. 10.3. Dynamometer in hobber.

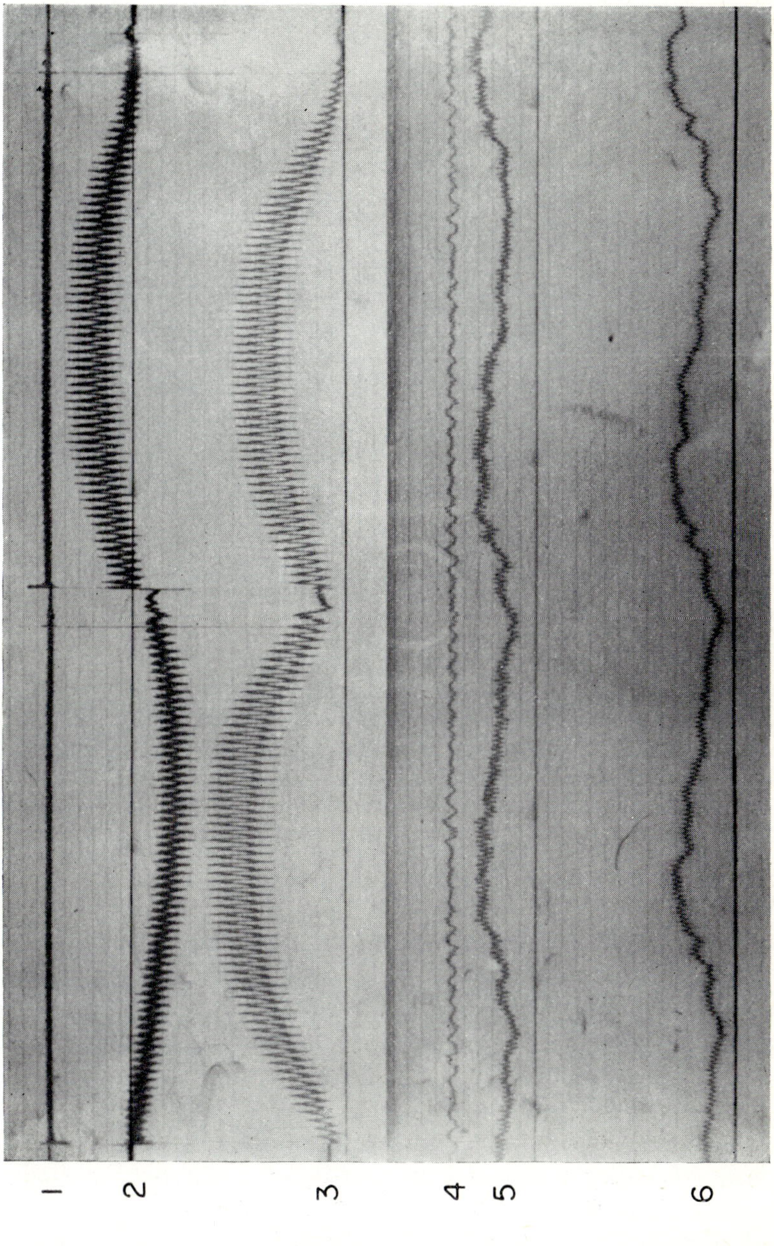

Fig. 10.4. Shaving forces for gear.

constant as the cutter was passed across and back again. The reversal of the direction of the force in trace 2 is due to the reversal of the cutting direction. Each spike in traces 2 and 3 represents one revolution of the cutter. In traces 5 and 6 each small spike represents one tooth of the cutter, whereas the periodicity represents one revolution of the cutter.

For some very low force applications such as hand lapping, Perspex may be used for the dynamometer, giving a sensitivity which is an order higher.

4. Vibration measurements. The important variable in machine-tool structures under given load conditions is the relative displacement between cutting tool and workpiece since this controls machine accuracy and metal cutting capacity. Stress in itself is unimportant as stresses in machine tools are generally extremely low; a stress as low as one ton per square inch would still give several thousandths of an inch deflection on a typical machine-tool structure, which may have a stiffness of 500 000 lb/in.

The primary measurement in cutting processes is the dynamic response between tool and workpiece. This is usually determined under test conditions using a vibrator to cover the frequency range of interest to produce a machine resonance curve. When the frequencies of the resonances have been established, the deflected shape of the machine is required to allow the contributions of the separate parts of the machine to the total deflection to be assessed.

5. The pick-ups available for machine investigation fall conveniently into these classes:

a. Relative displacement pick-ups: Contact. These usually work inductively from a relatively low frequency (8 kHz) a.c. carrier. Movement of a permeable rod alters the inductive coupling of coils and a conventional a.c. bridge will then give a signal proportional to displacement. They are very good for tool-workpiece deflection measurements when the machine is not running, and since their mass is small, they do not significantly alter resonance frequencies.

b. Relative displacement pick-up: Non-contact. These have a wide frequency range and are particularly suitable for model testing where pick-up masses must be negligible. They are usually inconvenient to set up and not easily movable.

c. Accelerometers. Since these measure acceleration the output usually needs a double integration to give a direct reading of displacement and to filter out the high frequency noise. Since the mass of an accelerometer is only a few ounces, it can be attached to a machine structure using a small pot magnet; the small additional inertia will not significantly affect the machine performance. When relative displacement between tool and workpiece is being measured, two accelerometers, two charge amplifiers and, if necessary, integrating amplifiers are required. This method, though expensive, is very convenient experimentally; accelerometers can now be calibrated reasonably easily with modern charge amplifiers which can operate d.c. for calibration.
d. Seismic pick-ups: Velocity output. A seismic mass is supported on springs and relative motion occurs between the mass which remains stationary and the outer case which is held against the machine tool. This relative motion is measured by making the mass a magnet and fitting a coil in the housing. The voltage from the coil is directly proportional to the relative velocity between coil and magnet. This type of pick-up is cheap and convenient but its mass (roughly 2 lb) may affect the resonance that is being measured and since its natural frequency is of the order of 10 Hz, low frequency measurements cannot be made. The physical size of the pick-up can be a disadvantage in cramped conditions.
e. Relative velocity pick-ups. These are similar to (*d*) but measure relative velocity between the case which must be held stationary and the central coil which is spring loaded into contact with the machine. The difficulties associated with holding the outer case steady restricts the use of this type of pick-up.

6. Vibration excitation. For machine-tool excitation the practical choice is between moving coil electromagnetic vibrators, electromagnetic pull vibrators and hydraulic vibrators. Electromagnetic pull vibrators are in practice expensive, limited in power, inconvenient to use and suffer from the major disadvantage that they can only pull and not push.

7. Moving coil electromagnetic vibrators. These are the simplest and easiest to use and have a very wide frequency range up to several kilohertz. Sizes range from 1 lbf to 50 lbf for portable vibrators. Though much larger ones are made, their size, weight and expense rule them out for machine-tool work. Their disadvantages lie in the fragility of the coil mounting and their low force levels, which allow representative testing of grinding machines but not of normal metal cutting machines.

8. Hydraulic vibrators. These employ an electrohydraulic valve, as used in aircraft servo controls, to direct pressure on to either side of a double acting hydraulic ram. A typical design will give $\pm$1 500 lbf peak up to 300 Hz, then a steady reduction of force down to about $\pm$300 lbf peak at 1 000 Hz when testing a good machine of high stiffness. Larger forces can easily be obtained but with a reduced frequency range. The resulting vibrator can be physically small, e.g. about $7\times3\times2$ in and much cheaper than a

Fig. 10.5. Vibrator in machine.

large electromagnetic vibrator. The additional facility of simple independent control of the mean force level, by means of a d.c. signal to the control coils, is useful for detailed testing. Fig. 10.5 shows such a vibrator designed by one of the authors (JDS) in use to excite a gear shaping machine along the line of the resulting cutting force and Fig. 10.6 shows a cross section.

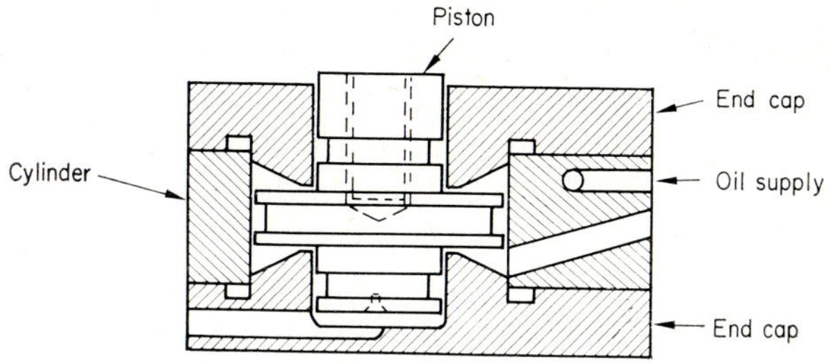

Fig. 10.6. Section of vibrator.

9. Too much emphasis cannot be placed on the importance of exciting machine tools at the force levels at which they will be working, with the force applied in the correct direction. Reliance cannot be placed on cross-receptance techniques in systems containing elements as non-linear as slides and joints in machine tools.

10. Problems

1. Is it better to measure displacement, velocity or acceleration when investigating:

(*a*) Low frequency vibrations when fast running mechanisms are giving high frequency noise?

(*b*) High frequency small amplitude vibrations?

(*c*) Low frequency vibrations of a large machine?

2. If a structure must be vibrated $\pm 0{\cdot}001$ in (25 μ) at the point of application of a vibrator what are the relative sizes and weights of an electromagnetic and electrohydraulic vibrator:

(*a*) If the structural stiffness is 2×10^4 lbf/in?

(*b*) If the structural stiffness is 5×10^5 lbf/in?

3. If a mass of 1 000 lb is sitting on an incline of 10° with a friction coefficient of 0·1 restrained by a spring of stiffness 10^4 lbf/in, what is the cross receptance for:

(*a*) Horizontal displacement with a vertical force of 200 lbf?

(*b*) Vertical displacement with a horizontal force of 200 lbf?

Appendix I

Elementary Phase-Plane Analysis

1. Information about the behaviour of systems is usually presented as a graph representing position plotted against time. For steady state sinusoidal motion information is also often presented in the form of Nyquist and other diagrams which give information about the relationship between input and output at varying frequencies. These loci are built up from vector diagrams which represent the relationships between the terms present in the equation of motion for the system, and these terms are themselves related to the displacement, velocity, acceleration and higher derivatives of the motion with respect to time. A great deal of information may sometimes be obtained about a system if attention is concentrated only on a plot of velocity against displacement; the diagram that results is known as the *phase portrait* of the system, and is in the *phase plane*. The reason for this name will become clear later.

2. Analysis of undamped spring-mass systems under varying force. If the force is given as a graph in terms of time, the curve may be broken down approximately into a series of step functions as in

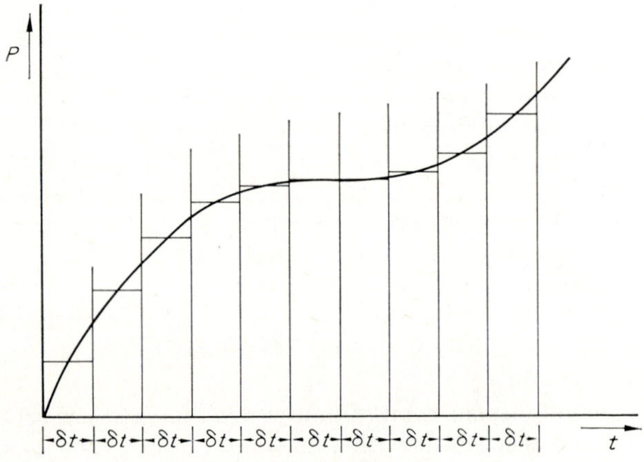

Fig. AI.1. Applied force.

Fig. AI.1. The behaviour of the mass during each step may be determined from elementary theory provided that the initial conditions are known: but to do this analytically is wearisome, and a graphical short cut may be sought.

Suppose that the system is as in Fig. AI.2, where the force P is applied to the mass m, its displacement from its mean position being x. For negligible damping,

$$m\frac{d^2x}{dt^2}+kx = P$$

$$\therefore \left[\frac{D^2}{\omega_0^2}+1\right]x = \frac{P}{k} \quad \text{where} \quad \omega_0^2 = \frac{k}{m} \tag{AI.1}$$

The solution of this equation is of the form, with A and ϕ as constants,

$$x = +\frac{P}{k} - A\cos(\omega_0 t+\phi) \tag{AI.2}$$

and differentiating to obtain the velocity

$$Dx = +A\omega_0 \sin(\omega_0 t+\phi)$$

Fig. AI.2. Spring-mass system.

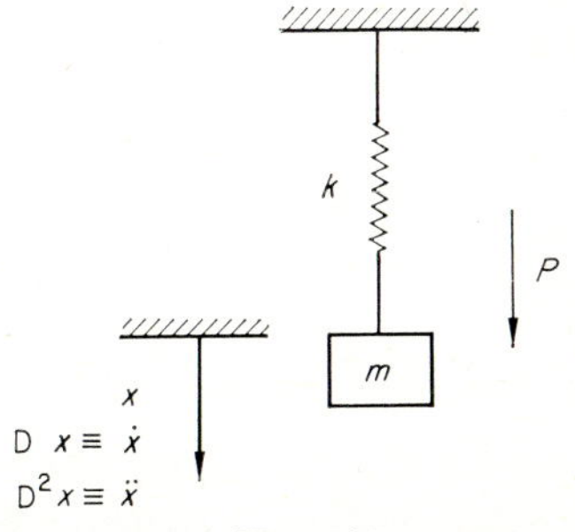

These equations are true for any step in Fig. AI.1. Taking any step, and putting at the commencement of that step $t = 0$, $x = x_0$, $Dx = \dot{x}_0$, the constants for the equations *during this step* may be found by substitution to give:

$$\frac{P}{k} - x_0 = A\cos\phi$$

and

$$\dot{x}_0 = A\omega_0 \sin\phi$$

Dividing gives

$$\cot\phi = \frac{\dfrac{P}{k} - x_0}{\dfrac{\dot{x}_0}{\omega_0}} \tag{AI.3}$$

Note that both top and bottom lines in the RHS of this equation have been arranged to have dimensions of length. Squaring and adding the equations gives

$$A^2(\sin^2\phi+\cos^2\phi) = \left(\frac{\dot{x}_0}{\omega_0}\right)^2+\left(x_0-\frac{P}{k}\right)^2$$

$$\therefore A = \sqrt{\left(\frac{\dot{x}}{\omega_0}\right)^2+\left(\frac{P}{k}-x_0\right)^2}. \qquad \text{(AI.4)}$$

These results may be represented graphically in Fig. AI.3 where the point $\left(\frac{\dot{x}}{\omega_0}, x_0\right)$ gives the initial conditions. At any time t later during this step, the equation is satisfied by a point on the *trajectory*, which is an arc of a circle with centre $\left(0, \frac{P}{k}\right)$.

At the end of time δt the trajectory will be at $\left(\frac{\dot{x}_1}{\omega_0}, x_1\right)$, thus giving the initial conditions for the next step. The angles on the figure show the changes in phase angle, and this is the reason for calling this representation the phase plane.

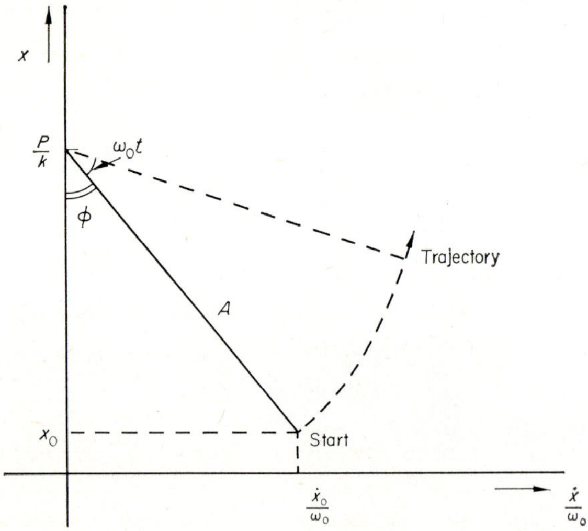

Fig. AI.3. Phase plane trajectory for a single step. $x = \frac{P}{k} - A\cos(\omega_0 t+\phi)$,

$$\frac{\dot{x}}{\omega_0} = +A\sin(\omega_0 t+\phi),\quad \cot\phi = \frac{\frac{P}{k}-x_0}{\frac{\dot{x}_0}{\omega_0}},\quad A = \sqrt{\left(\frac{\dot{x}_0}{\omega_0}\right)^2+\left(\frac{P}{k}-x_0\right)^2}$$

3. Special case of spring-mass system without damping or applied force. Since $P = 0$ in this case, the centre from which the arc is struck remains permanently at the origin, and the phase-portrait is a circle whose radius is determined by the amplitude of the initial motion.

4. Application of theory to a specific case. Because the forcing functions of interest are not ones which have simple expressions (although they could of course be expressed as Fourier series), general analysis is of little further help, and specific cases must be examined.

Fig. AI.4 shows an arbitrary force applied to an undamped spring-mass system. The mass is assumed to be initially at rest with no displacement from its mean position. If the mass is 40 lb and the stiffness of the massless spring is 2 lbf/in, find the subsequent motion of the mass. The steps in the solution are as follows:

a. Calculate $\omega_0 = \sqrt{\frac{k}{m}} = \sqrt{\frac{2 \times 32 \times 12}{40}} = 4{\cdot}37$ rad/s

$= 251°/\text{s}.$

b. Decide the time elements into which to divide the curve to give steps. Try $\delta t = 0{\cdot}1$ s. The accuracy of the procedure depends upon taking small enough time intervals: but if you rework the problem using $\delta t = 0{\cdot}05$ s and $\delta t = 0{\cdot}2$ s little difference will be found in the result. This is because the process performs a double integration graphically, and integration is essentially a smoothing process, as may be seen by integrating a Fourier series, where the terms of higher frequency rapidly diminish in importance.

c. Divide the curve into suitable steps.

d. Lay out phase plane in the position shown on the RHS of the figure. The scales are $\frac{\dot{y}}{4{\cdot}37}$ and $y = \frac{P}{k} = \frac{P}{2}$ in inches. Add this scale as shown to LHS of figure.

e. Locate initial value of $\frac{P}{k}$ in RHS to correspond with first step by projecting the step *1* to give the point *1a*.

f. With radius *1a* to origin, strike an arc of $251 \times 0{\cdot}1 = 25°$. This gives *1b*, which defines the end conditions of velocity and displacement for the first step; and the initial conditions for the second step.

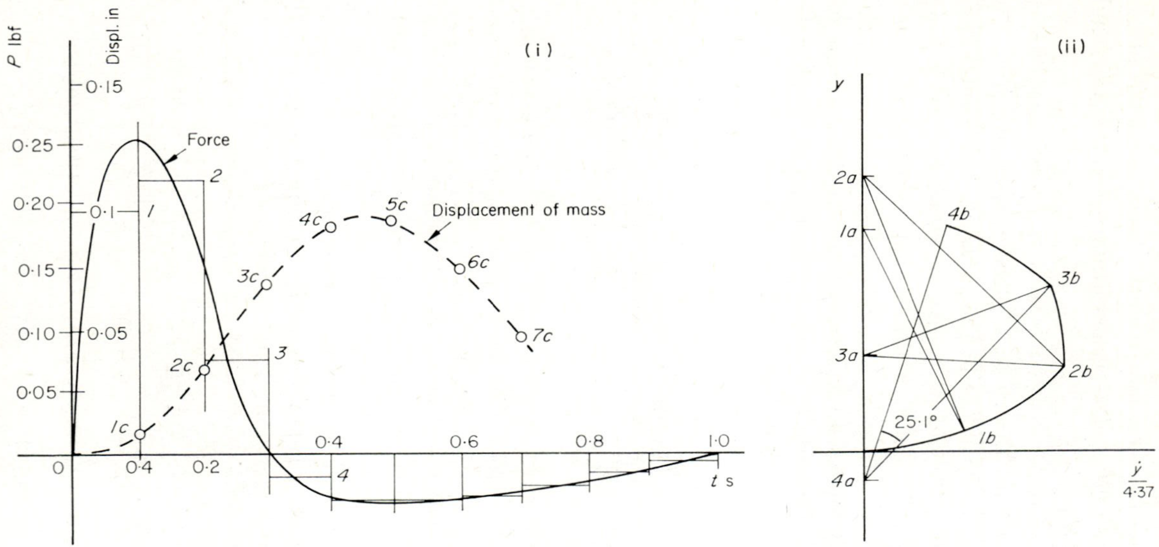

Fig. AI.4. Phase plane with arbitrary forcing. (i) Time Plane, (ii) Phase Plane.

g. Project point *1b* back to give *1c*, the corresponding point in the time plane at the end of the step considered.
h. Now repeat for steps *2*, *3*, *4*, etc., until the motion settles down to simple harmonic motion at time $t = 1$ s. Only the first few points have been shown in the drawing to avoid confusion.
i. The points *1c*, *2c*, etc., may be connected to display the resultant motion.

The convenience of plotting displacement vertically in the phase plane should now be apparent, since it enables results to be projected between the planes. The curves of displacement against time and against velocity may be regarded as two dimensional representations of a point moving in three dimensions.

5. Motion given, find the force applied. The inverse of this problem is the determination of the force-time curve if the ensuing motion is known. This construction is not so easy, and does not usually yield very accurate results, since it is essentially a process of differentiation. Details may be found of the method in a paper by Bishop.

6. Spring-mass systems with damping: viscous damping. If the damping is viscous, a simple construction may be found in Bishop's paper for allowing for it by the use of skewed axes and the use of logarithmic spirals instead of arcs of circles. However such, precision is rarely needed, and it may usually be treated with adequate accuracy by the general method given later.

7. Coulomb damping. The behaviour of a spring-mass system subject to a force of constant magnitude always opposed to the direction of motion is of considerable interest as an idealization of a more complicated real situation. The problem can be handled very easily with the aid of the phase plane.

Suppose that a force $P \sin \omega t$ is applied to a spring-mass system, and that the motion of the system is opposed by a force equal to $\frac{P}{4}$. This is typical of many elements in hunting systems. The equation of motion may be written

$$[m\mathrm{D}^2+k]y = P \sin \omega t \pm \frac{P}{4}$$

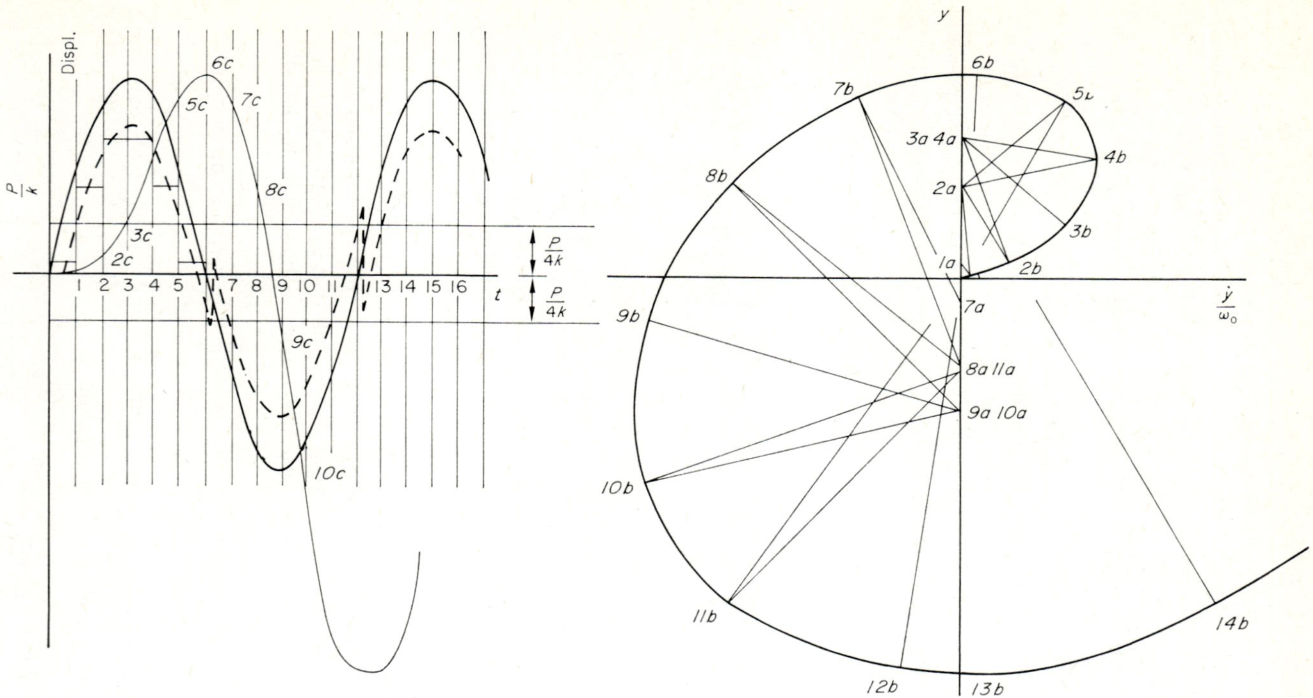

Fig. AI.5. Phase plane with Coulomb Friction.
Applied force ———; Net force - - - -; Displacement ———.

$\omega = \omega_0 = 2\text{rad/s}; t = 0; y = \dot{y} = 0; \delta t = \frac{1}{12}\text{ s}; \omega_0 \delta t = 30°.$

where + or − is to be taken so that the Coulomb force always opposes the motion. The method of solving this problem by phase-plane methods is shown in Fig. AI.5, for the particular case of forcing at resonance, when $\omega = \omega_0 = \sqrt{\frac{k}{m}}$. The construction can be carried out more rapidly if it is done on squared paper.

The steps in the construction are as follows:

a. Draw semi-circle on LHS and from it project to obtain sinusoidal input curve representing $\frac{P}{k} \sin \omega t$.

b. Insert lines of $\pm \frac{P}{4k}$

c. Construct the *first few points* only of net force curve. Note that until $\frac{P}{k} \sin \omega t > \frac{P}{4k}$ no movement can take place.

d. Construct phase-plane diagram and displacement curve for the mass as before until point *6c* is reached. Very shortly after this point the velocity of the mass will change sign, and therefore the curve of net applied force will show a discontinuity. Care must be taken in the construction at this point, since the mass is at rest, and will stay at rest so long as the net force curve remains $-^{ve}$.

e. Construct the next few points on the net input curve, and proceed as before until just after point *12b* when another discontinuity occurs. The mass will stay at rest until the net force becomes $+^{ve}$, and the point *13b* will approximately lie on the y axis.

f. Continue until the motion is 'steady', i.e. repeats itself.

If the chief interest in constructing such a curve is to find the steady state response, the work may be reduced by guessing initial conditions; in this case since the forcing and undamped resonant frequency are the same, a reasonable guess might have been to make $y_0 = \frac{P}{k}$, $\dot{y}_0 = 0$.

The mean phase lag between output and input can now be determined either from the phase-plane diagram or from the displacement curve. The displacement curve may also be subjected to Fourier analysis, either graphical or numerical, to ascertain the order of magnitude of higher harmonics.

8. The analysis may be much simplified if the undamped resonant frequency of the system is four or five times as high as the forcing frequency, since in this case the inertia force is only $\left(\frac{1}{4}\right)^2$ or $\left(\frac{1}{5}\right)^2$ the size of the spring force, and may conveniently be neglected. The construction then employed is shown in Fig. 7.24, and is no longer dependent on the phase plane.

9. Spring-mass system with any form of damping. In the previous section, the method of carrying out the analysis of a spring-mass system with Coulomb damping was demonstrated. Any form of damping may however be handled so long as its dependence on velocity and displacement are known. If, for instance, the damping depends on the velocity or upon the square of the velocity, then the velocity at the beginning of any time δt may be used to calculate a correction to the effective input force, and so to obtain the correct centre on the y axis from which to strike the arc in the phase plane. A better correction may often be obtained by estimating by eye what the average velocity during the ensuing time δt is likely to be.

Appendix II

Stability with three Simple Time Delays

The equation of a loop with three simple time delays, and loop gain μ, where θ is a variable, may be written

$$[T_1D+1][T_2D+1][T_3D+1]\theta+\mu\theta = 0 \qquad \text{(AII.1)}$$

$$\therefore [T_1T_2T_3D^3+(T_1T_2+T_2T_3+T_3T_1)D^2+ (T_1+T_2+T_3)D+(1+\mu)]\theta = 0$$

The condition for stability, using Routh's criterion, is

$$(T_1T_2+T_2T_3+T_3T_1)(T_1+T_2+T_3)>T_1T_2T_3(1+\mu)$$

$$\therefore T_1^2T_2+3T_1T_2T_3+T_1^2T_3+T_1T_2^2+T_2^2T_3+T_2T_3^2+T_3^2T_1> T_1T_2T_3+\mu T_1T_2T_3$$

Dividing through by $T_1T_2T_3$ gives

$$\frac{T_1}{T_3}+\frac{T_1}{T_2}+\frac{T_2}{T_3}+\frac{T_2}{T_1}+\frac{T_3}{T_1}+\frac{T_3}{T_2}>(\mu-2)$$

Suppose that T_2 and T_3 are invariable, but that T_1 may be altered: rearranging

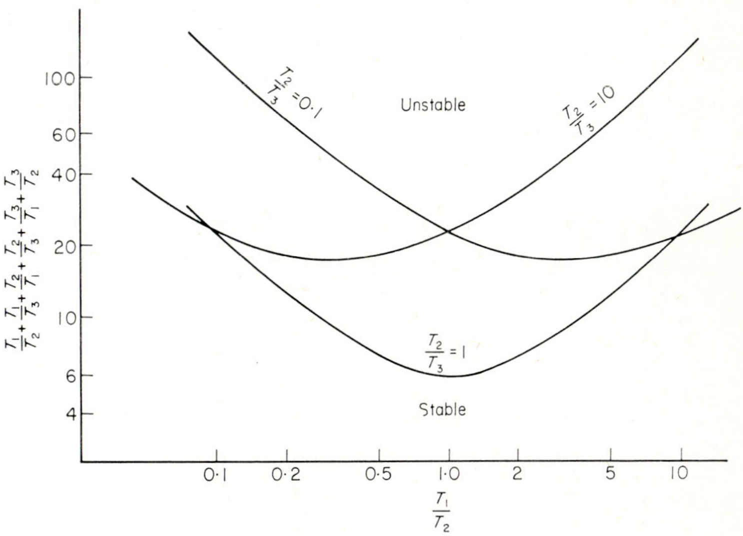

Fig. AII.1. Three lag stability boundaries.

$$T_1\left(\frac{1}{T_3}+\frac{1}{T_2}\right)+\frac{1}{T_1}(T_2+T_3)+\left(\frac{T_2}{T_3}+\frac{T_3}{T_2}\right)>(\mu-2) \qquad \text{(AII.2)}$$

If the LHS then has a minimum, i.e. $\frac{\delta}{\delta T_1}=0$

$$\left(\frac{1}{T_3}+\frac{1}{T_2}\right)-\frac{1}{T_1^2}(T_2+T_3)=0$$

$$\therefore\ T_1^2=T_2T_3 \qquad \text{(AII.3)}$$

If the value of the LHS of Eq. AII.2 is plotted against T_1 as a variable, a curve as shown in Fig. AII.1 will be found. For any given value of $(\mu-2)$ an unstable system may be stabilized *either* by making T_1 very small *or* by making T_1 very large.

Appendix III

Conversion Tables for Units

Length

1 ft	= 0·304 8 m	
1 in	= 25·4 mm	
0·001 in	= 25·4 μm	

Mass

1 ton	= 1 016·05 kg	
1 lb	= 0·453 6 kg	

Force

1 tonf	= 9·964 kN	= 1 016·05 kp
1 lbf	= 4·448 N	= 0·453 6 kp

(Mnemonic: 1 apple weighs roughly 1 Newton')

Torque

1 tonf ft	= 3·037 kN m	
1 lbf in	= 0·113 N m	

Stress

1 tonf/in^2	= 15·444 MN/m^2	= 157·448 kp/cm^2
1 lbf/in^2	= 6 894·76 N/m^2	= 0·070 3 kp/cm^2

Viscosity

1 cP	= 0·001 Ns/m^2	

Work

1 hp h	= 2·684 MJ	
1 kW h	= 3·6 MJ	
1 ft lbf	= 1·355 8 J	
1 therm	= 105·506 MJ	

Stiffness

1 lbf/in	= 175·127 N/m	= 17·858 0 kp/m

Bibliography

The following suggestions for further reading make no pretence at being more than a guide to some of the books and papers which the authors have found particularly useful.

Chapter 1. For a general background on machine tools, Professor F. Koenigsberger's *Design Principles of Metal-Cutting Machine Tools* (Pergamon 1964) is unique in the English literature. Professor S. A. Tobias's *Machine Tool Vibration* (Blackie 1965) was the pioneering book in the field of chatter. *The Essentials of Control Theory for Mechanical Engineers* (Arnold 1963) by D. B. Welbourn handles the essential control and vibration theory with the simplest possible mathematics. It is frequently referred to in this book as *ECT*, since the analytical approach is common to both. For papers on the application of more advanced analytical methods using statistical control theory, reference may be made in *Advances in Machine-Tool Design and Research: Proceedings of the International MTDR Conferences* (Pergamon) to papers by J. Bollinger, A. W. Kwiatkowski and others: a clear discussion of the principles is to be found in *Mathematics of Servomechanisms* by J. L. Douce (EUP 1963).

Chapter 2. Much useful work both on the statics and dynamics of machine-tool structures has been published privately for the use of members by the Machine Tool Industries Research Association, Macclesfield, and the Production Engineering Research Association, Melton Mowbray. The MTDR *Proceedings* should also be consulted, and in particular an article by F. M. Stansfield on model techniques in the *Proceedings* of the 6th Conference. The Russian Engineering Journal, which is available in translation, carries good work from 1959 onwards; in German, the *VDI Forschungsheft* Nos 409 and 429 are of particular use, as are also the FoKoMa proceedings.

Mechanical Vibrations (McGraw-Hill 1956) by J. P. den Hartog is a unique book which is an education in itself in the whole field of vibration analysis.

A short account of hysteretic damping may be found in *ECT*, Appendix I.

Chapter 3. Holzer's method is discussed in den Hartog's book. Papers on the transient characteristics of electrical machines are scattered through the literature. 'Linearised Analysis of Induction Motor Transients' by G. J. Rogers (*Proc. IEE*, **112**, No. 10, October 1965) is good.

Chapter 4. The literature on dynamic cutting forces tends to be unsatisfactory; P. W. Wallace and C. Andrew review it, and add useful work of their own in *J. Mech. Eng. Sci*. 1966, **8**, No. 2. A review is also given in J. D. Smith's Ph.D. Thesis 'Dynamic Metal Cutting' (Cambridge University Engineering Laboratory, 1962). Slow motion films of the cutting process may be obtained from The College of Aeronautics, Cranfield.

Chapter 5. Professor Tobias's book reviews the literature. In German, attention should be paid to J. Tlusty and O. Danek *Selbsterregte Schwingungen an Werkzeugmaschinen* (VEB Verlag 1962). For the method of handling finite time delays used in Para. 4 see *ECT*, Chap. 2.18; for backlash and stiction see *ECT*, Appendix I, 6.2 and Appendix III. For grinding chatter, see the paper by R. S. Hahn and R. P. Lindsay given at the MTDR conference, 1969.

Chapter 6. A good introduction may be found in *Chartered Mechanical Engineer*, March 1965, by Professor J. Loxham. Papers in the MTDR *Proceedings*, 1967, by M. Polacek and Z. Vavra and by F. M. Stansfield, A. K. Kher and A. Cowley give a good idea of the practical design problems involved.

For gas bearings, *Gas Film Lubrication* (Wiley 1962) by W. A. Gross deals well with the theoretical side, while the practical design side is well handled in *Gas Lubricated Bearings* (Butterworth 1964) by N. S. Grassam and J. W. Powell.

Chapter 7. *The I.Mech.E. Proceedings of the Symposium on Automatic Control*, 1960, contains much useful information; note in particular a paper by G. D. S. Maclellan, A. E. Mitchell and D. E. Turnbull, 'Investigation into the Accuracy of Hydraulic Servo-motors'. *Phillips Research Reports* 16, *Dec*. 1961 (*Reprint R* 442) by T. J. Viersma is also stimulating. Some of the analysis of this chapter is to be found in greater detail in *ECT*.

Chapter 8. When faced with difficult vibration problems, den Hartog's book will often bring inspiration. For those who prefer a matrix approach to vibration theory, *Mechanical Vibrations* (Longmans 1963) by J. M. Prentis and F. A. Leckie offers a good introduction.

An excellent book by a member of the staff of the Cincinnatti Milling Machine Company is *Hydraulic Control Systems* by H. E. Merritt (Wiley 1967). *Design of Hydraulic Control Systems* by E. Lewis and H. Stern (McGraw-Hill 1962) is also useful.

Chapter 9. Very many standard textbooks on mechanics of machines are misleading, because the thinking behind them is essentially 'semi-static'. *Mechanics of Machines* (Longmans 1964) by J. L. M. Morrison and B. Crossland has some useful things to say on balancing.

Chapter 10. Textbooks tend to be out of date as soon as they are published in this field. The most instructive reading is manufacturers catalogues; at the time of writing (1969) most essential equipment is covered by Southern Instruments (Camberley, England), Vibrometer Corporation (Fribourg, Switzerland) and Phillips (Eindhoven, Holland).

Appendix I. This topic is now frequently handled in an advanced manner in textbooks; a particularly enlightening paper is an early one, that by Professor R. E. D. Bishop in *Proc. I. Mech. E.* 1954, **168,** No. 10.

Appendix II. *'Control' Data and Method Sheet* 36 (November 1965) by R. D. Tyler and B. J. Cooper approaches this problem in a slightly different manner.

Appendix III. *Changing to the Metric System* (HMSO 1967) discusses British, Metric and SI units.

Index